T0205526

Loudspeaker Physics and Forced Vibration

William H. Watkins

Loudspeaker Physics
and Forced Vibration

 Springer

William H. Watkins
Kingsport, TN, USA

ISBN 978-3-030-91636-7   ISBN 978-3-030-91634-3   (eBook)
https://doi.org/10.1007/978-3-030-91634-3

Cover Image: Speaker design © Adriana R. Watkins

This Springer imprint is published by the registered company Springer Nature Switzerland AG
The registered company address is: Gewerbestrasse 11, 6330 Cham, Switzerland

Preface

Without forced vibration the world would be a very different place. Our vocal chords would emit no sound, and there would be no verbal communication. There would be no sound from your TV, we would hear no thunder during a thunderstorm, and musical instruments would produce no sound.

During my 52 years of working with dynamic loudspeakers and forced vibration, several interesting phenomena were encountered and various new relationships and expressions were developed. They serve to clarify the energy conversion occurring in a loudspeaker and the workings of forced vibration. It was felt that these observations might be of interest to others, hence this book.

The book is different from the usual stark technical tone of most technical books. It is written in "story-book" form, such that the function of each parameter of a reference dynamic loudspeaker (a device most everyone is familiar with) is analyzed. Its value is then calculated, and then measured to illustrate all facets of loudspeaker operation and forced vibration. Thus, the learning is made easier and more interesting. Note the principles explained herein apply to all direct reciprocating motors, not just those in a dynamic loudspeaker.

Groundwork is laid with discussion of analogies between electrical, mechanical, and acoustic properties, along with their impedance relationships. All aspects of loudspeakers and forced vibration are then discussed. These include energy conversion, motion, force, velocity, acceleration, mass effects, work done, vibration amplitude, back-EMF, momentum, SPL, efficiency, mechanical power, frequency response, acoustic power, bass extension, and both electrical and mechanical resonance considerations. All these parameters are discussed in detail, and for further clarification 33 graphs of parameter amplitudes across the frequency range and 10 tables of predicted and measured magnitudes are included. Also given is a simple and very precise method for the measurement of a loudspeaker's motor strength (Bl), and a technique to circumvent a fundamental limitation inherent in conventional loudspeaker systems.

Unique to the book is an entire chapter dedicated to the discussion of back-EMF voltage. Although it is the factor responsible for the actual transfer of energy in an electric motor, it is not generally well understood, and is usually glossed over or given short-shrift in the literature on motors. Its function is discussed from several different points of view and is then analyzed in depth as related to the transfer of energy, Newton's laws, magnetic force, conservation of energy, mechanical power, and drive voltage.

The author wishes to express his appreciation to the many writers who have gone before, shared their knowledge, and helped pave the way for this book. This is especially so for those listed in the references. Special thanks are also due to Stephen F. Smith for his assistance with the final editing.

Kingsport, TN, USA   William H. Watkins

Acknowledgment

The family of Bill Watkins would like to thank Dr. Stephen F. Smith, one of Bill's oldest and dearest friends, for working diligently with the family to bring this book to publication. It was through Stephen's attention to detail and many past conversations with Bill that he was able to complete the publisher's questionnaires, as well as review and provide final edits that were significant in making this important work available.

About the Author

William H. Watkins (1929–2018), was a true American audio innovator. He invented the famous Watkins Dual-Drive Woofer, the renowned WE-1 large floor-standing speaker system, the critically acclaimed compact Watkins Generation Four book-shelf loudspeaker, and numerous other high-performance speakers. He also wrote and sold an extensive ESL (English as a Second Language) database to Franklin Electronic Publishers and, at the time of his passing, had just completed a new ESL app.

Bill was born in 1929 in Kingsport, TN. His love of music came naturally—in his youth, he became an accomplished musician, especially on the mandolin, guitar, piano, and violin. As a teenager and young adult, he enjoyed performing and entered numerous music competitions in North Carolina, Tennessee, and Virginia, often winning prizes for his proficiency. At one point in his musical career, he played a set with famous folk artists Doc and Merle Watson in a North Carolina club.

He served in active duty in the U.S. Army in the 101st Airborne Division ("The Screaming Eagles") during the Korean War era. During that time, he and his wife lived in Trieste, Italy. Upon returning to the States in the late 1950s, Bill opened a successful food distribution service in Kingsport. Audio was always a hobby for Bill, and a little later he started a second business selling Hi-Fi stereo components. As the hi-fi business became successful, he sold the food distribution service and established Watkins Stereo, a retail and custom audio firm, and Watkins Engineering, which manufactured and marketed several of his later advanced loudspeaker systems, including the highly regarded WE-1 and Generation Four models.

Bill was granted three U.S. patents in the area of acoustics, including U.S. #3,838,216 in 1974 for the Dual-Drive Woofer concept, which significantly improved low-end performance of an acoustic-suspension system by effectively canceling the low-frequency resonance impedance peak, thus extending the bass frequency response by close to an octave.

He also authored a landmark article on the technology, "New Loudspeaker with Extended Bass," in the December 1974 edition of *Audio* magazine. The patented concept was subsequently licensed to Infinity Systems in California (at the time one of the three largest loudspeaker manufacturers in America) for use in many of their high-end speaker systems and soon it became widely known as the "Infinity/ Watkins" Woofer.

In 1982 Watkins collaborated with his son Bill Jr. on a second patent, U.S. #4,443,744, on reviewer-acclaimed sound-damping panels for home listening use, later marketed as the Watkins Echo Muffs. A third patent, on resistively damped crossover networks to optimize speaker transient response, U.S. #4,475,233, was issued to him in 1984.

His prolific speaker design career continued into the twenty-first century, culminating in the current highly reviewed compact Generation Four model, a two-way system incorporating multiple patent-pending features to assure unusually wide, smooth frequency response, along with precise damping and stereo imaging from a small ($\frac{1}{2}$ ft^3) enclosure.

Bill also ventured outside the world of acoustics to write two ESL programs, one of which he sold to Franklin Electronic Publishers for use in their hand-held electronic translator devices in the 1980s. At the time of his passing, he had just completed writing a new, advanced ESL app.

One of his favorite hobbies was listening to all genres of music (classical, blues, jazz and folk were his favorites) as his extensive vinyl and CD collection bears testimony. His music collection totaled more than 5000 records and thousands of compact discs. Many of the albums and CDs bear his personal notes on the sound reproduction as heard through his various loudspeakers.

Bill was a longtime member of the Audio Engineering Society, which he originally joined in 1974. He maintained an extensive library of AES publications, along with issues of *Audio, Stereophile* and *The Absolute Sound* magazines over a span of several decades, which he heavily utilized in his ongoing audio and electroacoustics research and development activities.

He was married for 68 years to his devoted wife, Nell Walters Watkins; and left behind his son, William D. (Bill Jr.) Watkins and his wife Melissa; his daughter, Renee W. Dean, two granddaughters, Adriana Watkins, and Audrey Altonen and her husband Tim, and their daughter Adeline.

Truly his wit, rare insights, and deep knowledge of sound and music will be sorely missed by those in the audio field who knew him. The Watkins company, under Bill Jr., is continuing in the Hi-Fi audio business in Kingsport; their website is www.watkinsaudio.com, and the email address is watkinsaudio@gmail.com.

Introduction to *Loudspeaker Physics and Forced Vibration*

This book from internationally known loudspeaker and acoustics designer William H. Watkins presents a comprehensive and highly insightful study into the workings of dynamic loudspeakers and dynamically forced vibration. The fundamental parameters of a typical dynamic loudspeaker, along with actual measured values, are utilized in the performance equations throughout the work as illustrative examples of all facets of forced vibration in general and loudspeakers in particular. This presentation style makes the analysis more engaging, interesting, and intuitive to comprehend compared to most previous works in the field.

The book is based on a lifetime of research and widely acknowledged commercial loudspeaker design successes by acclaimed speaker designer Bill Watkins. The text differs from the usual stark technical tone of most technical books on this subject matter. The book's style is to initially present and then fully analyze the function of each fundamental parameter of a reference dynamic loudspeaker. Its value is then calculated and also confirmed via lab measurements to vividly illustrate all energy-transduction facets of loudspeaker operation and forced vibration.

The principles of this book apply to all direct reciprocating motors, not just those in a dynamic loudspeaker. Unique to the book is an entire chapter dedicated to the discussion of back-EMF voltage, discussed from several technical points of view and analyzed in depth as related to the dynamic transfer of energy within the loudspeaker unit between the mechanical and electrical domains. Another key feature is a detailed discussion of Watkins' patented "dual-motor" concept to achieve high dynamic speaker performance in the region of its low-frequency resonance.

A concise, chapter-by-chapter presentation of the essential topics provides a clear understanding of the physics behind both forced vibration in general and dynamic loudspeaker transducers in particular. This includes numerous properties rarely covered by previous texts, with all the salient parameters of operation being examined in a classic but readily understandable format. This work is replete with numerous comparisons of theoretical values and corresponding laboratory measurements. Also included is a glossary of key technical terms and exhaustive lists of pertinent references. The degree of thoroughness and detail contained in this very readable work has not been available previously in earlier sources.

This book provides a unique perspective of the detailed physics of the energy balance and electromagnetic energy-conversion processes within the loudspeaker itself not found in any other text. Key features of the book include first providing a comprehensive first-principles physics-based analysis of dynamic loudspeaker operation. The various parametric analyses are then augmented with confirming high accuracy laboratory measurements. Most of these measurements are then also cross-checked and confirmed with multiple complementary measurements of related electroacoustic properties of the test driver. Full correlations of these results to those found in other classic acoustics texts and references are also provided. Finally, a unique, detailed explanation of the renowned Watkins Dual-Motor concept (U.S. Patent 3,838,216) is included in Chapter 16.

The text begins in Chapter 1 with a brief recitation of the key technical terms used throughout the book, which gives an early introduction to the scope of the work and provides a curiosity-boosting inducement to press on into the details of the physics of dynamic loudspeakers. The chapter then immediately jumps into the "meat" of the energy-conversion and motional processes within the dynamic loudspeaker. Chapter 2 launches into electroacoustic measurements, electrical/mechanical/acoustical analogies, and the development of the new power-to-loss ratio and volume acceleration parameters. Chapter 3 presents a succinct view of the equivalent electrical circuit of the driver in a closed-box environment, incorporating the concurrent electrical, mechanical, and acoustical effects on the loudspeaker system. Specific values in the representative circuit for the reference driver are also provided.

Chapter 4 first enumerates the five sources of equivalent resistance in the operating driver: (1) voice-coil DC resistance; resistance due to (2) mechanical and (3) acoustic loads; (4) magnetic core resistance due to eddy-current flow; and (5) effective source resistance in the amplifier and wiring. A discussion of blocked-coil laboratory measurements follows, along with considerations of both static and motion-induced reactances, driver inductance factors, overall system impedance measurements, and the development of comprehensive input voltage/power expressions. Next, Chapter 5 begins with analyses of overall efficiency and the novel concept of the loaded driver's mechanical response function. Subsequent sections deal with the relation of these parameters to driver input power and correlation of these expressions to earlier texts including those by Beranek, Kinsler, Kloss, and Small. Finally, cycle-by-cycle power and energy calculations are examined.

Chapter 6, focusing on the often-obscure topic of "back-EMF," opens with the application of Faraday's and Lenz's laws, proceeding to the transfer of energy in the driver, a detailed analysis of the processes producing back-EMF, and the application of Newton's laws of motion for the driver. The chapter closes with calculations of energy conservation and a thorough analysis of the intimate relation between back-EMF and power transfer. Chapter 7 examines details of the mechanical and acoustic load impedances, including air loading on the moving cone.

Chapter 8 provides a discussion of mechanical and acoustic power parameters, including electrical and mechanical power factors and calculations of mechanical-plus-acoustical power. Next are offered acoustic power relations to kinetic energy effects, followed by the distribution of energy and power over the driver's operational AC cycle. Calculations of power losses and key driver parameters, including overall efficiencies as a function of drive frequencies, are also tabulated. Chapter 9 briefly introduces details of the driver's closed-box low-frequency response, mainly in the region around mechanical resonance. The concepts of the half-power point, system Q_T, and resonance frequency measurement techniques are also covered. (Many more details of the dynamic low-frequency system response are provided later in Chap. 16.)

Chapter 10 contains a comprehensive discussion of the forces on the driver's moving parts (cone and voice coil), including the Lorentz force law ($f = Bli$), motor resistance and electromagnetic damping, a correlation with Beranek's classic cone-velocity expression, and relations to the driver power-to-loss ratios presented earlier in the text. The chapter closes with an analysis of the key electromagnetic force factors and motional equations for the system under sinusoidal drive.

Chapter 11 focuses on the parameter of cone velocity in the closed-box system, providing discussions of average and RMS velocity under variable frequency drive conditions. Complete expressions for the relevant cone motions are provided, along with details of laboratory cone-motion measurements. The poorly understood but important concept of recoil velocity (i.e., the relative recoil of the driver's magnetic motor assembly in normal operation) is also presented, along with confirming laboratory measurements of the effect. Chapter 12 provides a useful analogy of the potential and kinetic energy of the motion of a wrecking ball to that of a speaker cone driven by a cyclic waveform and includes evaluations of the cone's kinetic energy and source power drawn as a function of drive frequency over the cycle.

Chapter 13, titled "Work, Efficiency, and Power," analyses the energy expended in heating the voice coil and that required to move the cone over a typical (sinusoidal) AC drive cycle. The work done on the cone in both mechanical and acoustical domains is both calculated and graphed as a function of drive frequency and compared with the coil heating and motional components of that work. Additionally, the measured relations of the drive power to the power-to-loss ratios versus frequency are presented. Chapter 14, "Resonance, Q, and Measurements," features discussions of the effects of cone mass, suspension/box compliance, and damping on the fundamental low-frequency resonance and the Q and bandwidth of that resonance. The analytical results are also closely correlated with precision lab measurements and the factors of electromagnetic and overall system damping, Q_E and Q_T, were compared. Chapter 15 provides a discussion of cone motion, including basic acceleration, the volume acceleration (cone-area dependent), and the cone-motion amplitude. From these, the SPL (sound pressure level) is calculated. Next, a significant analysis of the Bl (motor) parameter is detailed and related to the back-EMF. A careful first-principles lab measurement of the Bl force is then offered, along with a detailed technique required for accurate results. Finally, calculations of cone momentum and the resulting acoustic pressure level are presented.

Chapter 16, titled "Circumventing Efficiency and Bass Extension Limitations," perhaps the most fascinating material in the book, provides an in-depth discussion of the highly acclaimed Watkins Dual-Motor Woofer, which gained commercial fame in the 1970s and 1980s in multiple audiophile speaker systems produced both by the Watkins firm and also by Infinity Systems of Chatsworth, CA. The revolutionary technique was granted U.S. Patent 3,838,216 in 1974 and was further documented in a landmark article by the same name in *Audio* magazine in December 1974. Fundamentally, the concept permits the decoupling of the normally inherent tradeoff between the flat-band efficiency of an acoustic-suspension (closed-box) woofer and its relative efficiency in the region of the fundamental low-frequency resonance. The surprisingly simple solution is implemented by the use of a secondary voice coil on the woofer's former, fed from a series-resonant LC network tuned to the fundamental acoustic resonance of the driver in the box and driven from the same source as the main coil. For flattest response and bass extension, the Q of the secondary circuit must be matched to the dynamic response of the woofer in its enclosure. The key to understanding the invention lies in the behavior of the speaker's back-EMF, especially around the in-box cone resonance. The scheme so effectively reduces the typical peak in the back-EMF near resonance that the woofer's drive impedance appears to be largely resistive, thus coupling much better to most modern audio amplifiers and also providing appreciably improved transient response and lower bass distortion. The chapter also provides extensive performance measurements of the dual-motor system to further document the superiority of the approach over conventional woofers.

Finally, the book concludes with a summary of measured test driver parameters and tabular data versus frequency in Appendix A and a copy of the landmark Audio magazine article on the Dual-Motor concept in Appendix B. A detailed Glossary and Symbols section provides an exhaustive enumeration of the key technical terms used in the book, followed by a standard Reference section and Index of terms.

Overall, the book assumes only a basic working knowledge of simple electrical circuits and classical Newtonian physics and could be gainfully utilized in a university electrical engineering or physics curriculum as an advanced undergraduate or even graduate-level text. With its wealth of practical observations and real-life examples, the work will also prove invaluable to the practicing motor design or loudspeaker design engineer.

The work herein is based on a lifetime of intense research, experimentation, and laboratory measurements by Mr. Watkins. The final result of all this effort has been the critical acclaim and commercial success of his many groundbreaking designs, including the Watkins WS-1, WS-1a, WE-1, and Generation-4 loudspeakers, the Infinity-Watkins woofer subsystems (embedded in the Infinity Quantum 2, 3, 4, and 5 units, the QLS, QRS, Infintesimal, RS 2.5 and 4.5 models, and the Renaissance 80 and 90 systems), and the room-acoustics enhancing Watkins Echo Muffs. I highly recommend this uniquely insightful book, which contains a veritable treasure-trove of useful information and perspectives to both assist and inspire future designers in advancing the state of the loudspeaker art and motor design.

Stephen F. Smith, Ph.D. (EE)
Senior R&D Staff Member (Retired)
Oak Ridge National Laboratory,
Oak Ridge, TN
Currently: Consultant to broadcast &
communications industries

Contents

Chapter 1
Prologue

1.1 Abstract

A direct-radiator loudspeaker is analyzed to illustrate all aspects of the physics of forced vibration. The electrical, magnetic, mechanical, and acoustic dimensions of loudspeaker operation are examined, and a worked example of a reference loudspeakers parameters is used throughout the book. This serves to illustrate in a simple manner all the various aspects of forced vibration, as well as the energy conversion occurring in a loudspeaker. As noted in the preface, the principles explained apply to all direct reciprocating motors, not just those in a loudspeaker.

A full derivation is given for the inductive component of the electrical input impedance, the resistance of the voice coil, the core, and that of the mechanical and acoustic load as seen by the source. The calculated impedance is then shown to match measured impedance.

True efficiency is given as derived from true input power, and as related to f_3, volume acceleration, kinetic energy, work done, and the power-to-loss ratio. The true velocity with allowance for motor recoil is also discussed.

The motion generated back electromotive force (back EMF) is analyzed in detail, and is shown to be key to the amount of power transferred.

A derivation is given for electromagnetic damping $(Bl)^2/(R_E + R_{ECOR})$ due to the transducer's motor resistance. It is then shown that a dedicated force overcomes this damping effect. The mechanical stage of power transfer is analyzed and both mechanical and electrical power factors are discussed. Also, kinetic energy as related to input power is discussed.

The radiated acoustic power is given as related to mechanical power, kinetic energy, the power-to-loss ratio, and volume acceleration. Newton's force law $F = Ma$ is shown to be related to the major speaker parameters, and an overall response function allows the prediction of efficiency at all frequencies using simple measurable parameters. Further, several simple expressions for f_3 are given, along with equations of motion as related to force.

© The Author(s), under exclusive license to Springer Nature Switzerland AG 2022
W. H. Watkins, *Loudspeaker Physics and Forced Vibration*,
https://doi.org/10.1007/978-3-030-91634-3_1

1

All relevant parameters such as acceleration, amplitude, motor strength, etc. are discussed, and specific methods for accurate measurement of the basic parameters are given, and a simple-to-implement and inexpensive method to circumvent efficiency and bass extension limitations in a given size of loudspeaker enclosure is described.

Two terms new to this context, i.e. volume acceleration and the power-to-loss ratio are also presented. Finally, 33 graphs and 10 tables are presented to show the magnitudes of all major parameters across the pertinent frequency range.

1.2   Elaboration

The inductance of the voice coil does not behave as that of an ideal inductor [1–5]. It is generally modeled as a lossy inductance because of power losses in the resistive components of the core (mainly the pole piece) due to eddy current flow. Finding the core resistance, along with voice coil electrical resistance plus that due to the mechanical and acoustic load, i.e. the total system resistance, allows the true inductive component of the input impedance to be found and modeled in a simplistic manner. This analysis process is described and the impedance is then calculated and compared with the measured impedance. The knowledge of exact total resistance also allows determination of the true input power, from which true overall conversion efficiency is then calculated and analyzed in depth.

Back EMF (often denoted as lower-case "emf"), also known as counter-emf or speed voltage, is often mentioned in texts without any detailed discussion of its function. Its pivotal role in energy transfer is discussed in depth, along with key supporting equations. The overall power conversion process is described, including the distribution of the input electrical power into the voice coil, core, mechanical load, and acoustic load components.

The mechanical force-distribution process is also described in detail, where the total applied force is apportioned to overcome the retarding force of the motor, the mechanical load, the acoustic load, and that due to reactance.

The motor resistance is seen on the mechanical and acoustic side of the overall equivalent circuit as a damping resistance, calculated as $(Bl)^2/(R_E + R_{ECOR})$. A complete derivation for it is then offered, and its effect as a damping force is shown to require a dedicated complementary force to overcome. The motor damping and Q_T parameters largely determine the half-power frequency f_3 in the low end response roll-off. Along this line, several simple and intuitive expressions for f_3 are given.

In general, Newton's force law $F = Ma$ may not be applied directly here due to the effects of mechanical resonance. Thus, a mechanical response function is described that compensates for this limitation and further provides a simple expression for efficiency. This expression is valid at all frequencies of transducer operation.

An explicit relationship is then given that allows calculation of kinetic energy from the applied power.

A recoil force is generated as the coil and cone unit is propelled back and forth by the alternating drive current. This force is in the direction opposite the coil-cone movement. It affects the measured velocity, and is addressed along with the direct measurement of recoil velocity.

Finally, two concepts new to loudspeaker analysis and forced vibration, volume acceleration and the power-to-loss ratio, are presented and elucidated. Their usage allows the simple calculation of many basic loudspeaker functions and forced vibration parameters, as will be seen.

Chapter 2
Preliminaries

2.1 General Measurement Conditions and Procedures

The parameters of twelve different high quality 8-inch drivers were evaluated with emphasis on linearity, low distortion, and stability. From these a Vifa model P21WN-20 was selected as the reference driver, and its specific measurements are used in the equations presented throughout this text. The unit was installed in a very rigid closed box with 1.25-inch thick laminated walls and no stuffing. The internal volume of the box was one cubic foot. Room temperature was maintained at 72° Fahrenheit, and the driver was driven with 1.41 volts *RMS* for a full hour before measurements were conducted. This insured stabilization of the voice coil and core temperatures, as well as the driver suspension's compliance. Each measurement was repeated several times to insure accuracy and consistency. These measured magnitudes are used in all equations throughout this work as illustrative examples to obtain parameter values. The applied voltage for all tests was exactly 1.41 volts *RMS*. The standard SI system of units (i.e., the MKS system) is employed, and all recorded electrical values are true *RMS* measurements (Fig. 2.1).

2.2 Assumptions and Limitations

The basic assumptions embodied in the driver measurements were as follows: (1) negligible source resistance (i.e., amplifier damping factor of 50 or greater); (2) radiation into a 180° solid angle (hemispheric) space, free-field, with the driver in a closed-box baffle; (3) mechanical resistance, compliance, and mass were assumed to be constant with frequency; (4) the mass of the moving assembly and that of the air load are considered to be combined, as is the reactance of the moving assembly and that of the air load; (5) small-signal and linear operation are assumed; and (6) the values given are for operation in the piston band at a reference frequency of 456 Hz,

© The Author(s), under exclusive license to Springer Nature Switzerland AG 2022
W. H. Watkins, *Loudspeaker Physics and Forced Vibration*,
https://doi.org/10.1007/978-3-030-91634-3_2

Fig. 2.1 Box used for measurements

unless otherwise noted. The data in some of the graphs from around 900 Hz and upward is affected by the cone not acting precisely as a piston, and thus may not follow the usual rules. Some analytical expressions may not apply above this range as well. To simplify concepts, in some cases the unipolar positive and negative values may be noted as equal in magnitude (symmetrical), i.e. the same "distance" away from zero. In other words, form favors clarity.

2.3 Electrical, Mechanical, and Acoustic Analogies

Developing an intuitive familiarity with the analogical system that relates electrical terms to mechanical and acoustic terms allows a clearer conception of mechanical and acoustic phenomena. The system treats mechanical and acoustic terms the same way mathematically as their analogous electrical counterparts. Table 2.1 provides a comprehensive listing of the key electrical, mechanical, and acoustic quantities used in this text. For example, current is analogous to velocity; therefore, just as electrical power is given as $P = I^2R$, then acoustic power is given as $P_A = v^2 R_{MA}$. The relationships shown in Table 2.1 are key to developing equivalent circuits for the transducer system, including driver, enclosure, and acoustical factors.

Table 2.1 Electrical, mechanical, and acoustic analogies

Electrical	Mechanical	Acoustic
Voltage	Force	Pressure
Current	Velocity	Volume velocity
Electrical resistance	Mechanical resistance	Acoustic resistance
Electrical impedance	Mechanical impedance	Acoustic impedance
Inductance	Mass	Acoustic mass
Capacitance	Compliance	Acoustic compliance
Charge	Displacement	Volume displacement
Power	Power	Power

2.4 Power-to-Loss Ratio and Volume Acceleration

Two terms new to this context are now developed which will be useful in the following discussions: the power-to-loss ratio PL and the volume acceleration Va. The power-to-loss ratio is the ratio of mechanical plus acoustic power to the dissipative power creating heat in the coil and core. In other words, this represents the ratio of usable output power (although in this case the mechanical portion is not directly useful) to the power lost as heat in the coil and core during the energy-conversion process. The overall power-to-loss ratio is given as

$$PL = \frac{I^2 R_{EMA}}{I^2 (R_E + R_{ECOR})} = \frac{R_{EMA}}{(R_E + R_{ECOR})} = 0.0156 \qquad (2.1)$$

or, alternatively, as the ratio of the effective portion of the back emf voltage (the component in phase with the current) to the portion of the applied voltage creating heat in the coil and core

$$PL = \frac{E_B \cos \phi_{MA}}{I (R_E + R_{ECOR})} = 0.0156 \qquad (2.2)$$

where

$$\phi_{MA} = arc \cos \left[(P_{MEC} + P_A)/F v \right] = arc \cos \left[R_{MEQ}/(Z_{MEC} + Z_A) \right]$$

or as related to resistances

$$PL = \frac{R_{ET}}{(R_E + R_{ECOR})} - 1 = 0.0156 \qquad (2.3)$$

The power-to-loss ratio may also be expressed as the ratio of work done per cycle on the combined mechanical and acoustic loads to that lost per cycle creating heat in the coil and core

$$PL = \frac{W_{EMACY}}{W_{EHCY}} = 0.0156 \tag{2.4}$$

or as the ratio of motor resistance to the combined mechanical and acoustic impedances

$$PL = \frac{R_{MTR}}{Z_{MEC} + Z_A} \cos \phi_{MA} = 0.0156 \tag{2.5}$$

The volume acceleration is the time rate of change of volume velocity. In *RMS* units at 456 Hz this is

$$Va = \frac{\Delta U}{\Delta t} \frac{0.5\pi}{\sqrt{2}} = \omega U = 1.4293 \tag{2.6}$$

or in other terms, just acceleration times cone area

$$Va = a S_D = 1.4293 \tag{2.7}$$

In terms of emitted acoustic pressure, this becomes

$$Va = \frac{2\pi}{\rho_O} pr = 1.4293 \tag{2.8}$$

and finally in terms of the new power-to-loss ratio we get

$$Va = \frac{\omega P S_D}{F \cos \phi_{MA}} \frac{PL}{PL + 1} = 1.4293 \tag{2.9}$$

The volume acceleration parameter has dimensions of L^3/T^2 and the SI unit is thus m^3/s^2.

2.5 Terminology

As is generally acknowledged in the electroacoustics literature, above mechanical resonance the system is said to be mass-controlled. This is due to the fact that in this region the driver's mass reactance is greater than its stiffness reactance. Below mechanical resonance the stiffness reactance of the suspensions dominate, and the system is said to be either compliance-controlled, where reactance $X_C = [1/(\omega C_{MT})] = 1.205$, or stiffness-controlled, where the reactance $X_C = (k/\omega) = 1.205$. Either phrasing is technically valid, and compliance or compliant reactance will generally be used here. As related to acoustic power, true efficiency is defined as the ratio of acoustic power radiated on the front side of the cone to the true electrical power delivered by the source. The motor metal assembly (magnet, pole piece, and plates) is referred to as the core.

Chapter 3
Analogous Electrical Circuit

3.1 Analogous Circuit and Discussion

The overall analogous equivalent circuit for the speaker system under consideration is shown in Fig. 3.1. The magnitude of the combined mechanical and acoustic impedances of the load is given in terms from the mechanical and acoustic side.

$$Z_{MEC} + Z_A = (R_{MS} + 2R_{MA}) + j\left(\omega M - \frac{1}{\omega C_{MT}}\right) = 72.886 \tag{3.1}$$

or as analogous to $Z = E/I$ [6]

$$Z_{MEC} + Z_A = \frac{Bll}{v} = 72.886 \tag{3.2}$$

This mechanical plus acoustic impedance is transferred through the motor's magnetic field into an electrical impedance

$$Z_{EMA} = \frac{(Bl)^2}{Z_{MEC} + Z_A} = 0.7331 \tag{3.3}$$

The source then sees the real part of this as

$$R_{EMA} = \frac{(Bl)^2}{Z_{MEC} + Z_A} \cos \phi_{MA} = 0.100 \tag{3.4}$$

This represents the resistance into which power is dissipated to drive the mechanical and acoustic load. See Sect. 4.2 for elaboration on R_{EMA}. Note that Eqs. (3.3) and (3.4) show the speaker to be an impedance inverter. A lower mechanical plus acoustic impedance translates to a higher value of R_{EMA} on the electrical side and

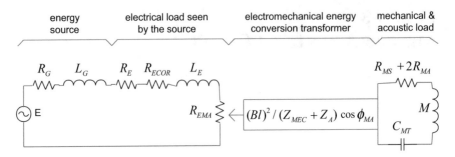

energy source | electrical load seen by the source | electromechanical energy conversion transformer | mechanical & acoustic load

Fig. 3.1 Analogous circuit for a moving coil loudspeaker in a closed box direct-radiator loudspeaker system. The resistive part of the mechanical and acoustic load is brought over to the electrical side. This is the resistance that source energy is dissipated into to drive the mechanical and acoustic load. Energy conversion from the electrical side to the mechanical and acoustic domain is accomplished in the magnetic field of the loudspeaker

reduces current flow. This is a reasonable expectation, since a lighter load requires less current to drive it. This process occurs at mechanical resonance and the reduced current flow reduces force Bl to match the lesser drive requirement of the load.

3.2 Reference Driver Magnitudes in the Analogous Circuit

Other magnitudes and functions in the analogous circuit are summarized here for the reader's convenience.

$E = 1.41 =$ input voltage
$R_G = 0 =$ resistance of the source, assumed negligible
$L_G = 0 =$ inductance of the source, assumed negligible
$R_E = 5.685 =$ resistance of the voice coil
$R_{ECOR} = 0.728 =$ resistance of the core
$L_E = 6.14 \times 10^{-4} =$ inductance of the coil, core, and that due to motion
$R_{EMA} = 0.100 =$ effective electrical resistance due to the combined mechanical and acoustic loads
$R_{MS} + (2R_{MA}) = 7.648 =$ resistance of the combined mechanical and acoustic loads
$M = 0.02572 =$ mass of the combined mechanical and acoustic loads
$C_{MT} = 2.896 \times 10^{-4} =$ compliance of the combined mechanical and acoustic loads
$(Bl)^2 = 53.436 =$ motor force constant (motor strength)
$\phi_{MA} = 82.164$ degrees $=$ mechanical + acoustic phase angle

Chapter 4
Electrical Impedance

4.1 Coil Resistance

The speaker's electrical impedance is a multi-faceted term. It consists of resistance from five sources, along with reactances from five sources as well. The five resistive elements are: (1) R_E, the DC resistance of the voice coil; (2, 3) R_{EMA}, the resistance due to the mechanical and acoustic load; (4) R_{ECOR}, the resistance of the magnetic core; and (5) R_G, the source resistance of the amplifier and speaker leads. In the following paragraphs, these resistive components will be examined in detail.

The initial term, R_E, is the resistance encountered by the source due to the voice coil alone; i.e., the DC resistance of the voice coil. The DC resistance R_E of the reference driver was measured with a calibrated DC source voltage as $R_E = E/I = 5.685$ ohms.

4.2 Resistance Due to the mechanical and Acoustic Load

Terms 2 and 3 are represented by R_{EMA}. This is an effective resistance seen by the source due to the combined mechanical and acoustic loads encountered by the driver.

The total resistance $R_{ET} = Z_{ET} \cos \phi_E = 6.513$, where $Z_{ET} = E/I = 6.7464$ ohms, and $\cos \phi_E = P/(E I) = 0.9654$, where voltage, current, and power were all explicitly measured. The core resistance is $R_{ECOR} = P_B - (I_B{}^2 R_E)/I_B{}^2 = 0.728$, where power and current measurements were again explicitly made. Now we see that coil plus core resistance is only $R_E + R_{ECOR} = 6.413$ ohms, leaving $R_{ET} - (R_E + R_{ECOR}) = 0.10$ ohm remaining that still must be accounted for. It in turn may be found as the real part of the mechanical plus acoustic impedances, after conversion to the electrical side as shown in Eq. (3.4), or as given by Kinsler [7]

© The Author(s), under exclusive license to Springer Nature Switzerland AG 2022
W. H. Watkins, *Loudspeaker Physics and Forced Vibration*,
https://doi.org/10.1007/978-3-030-91634-3_4

$$R_{EMA} = \frac{(Bl)^2 R_{MEQ}}{(Z_{MEC} + Z_A)^2} = 0.100 \tag{4.1}$$

The voltage drop across R_{EMA} is effectively the portion of the applied voltage driving the mechanical and acoustic load, which we shall call the drive voltage $E_{DR} = I R_{EMA}$. It follows that the electrical input power driving the mechanical and acoustic load is $I^2 R_{EMA}$. Power on the output side is thus given as the component of force doing work times velocity $F v \cos \phi_{MA}$, so that power out is equal to power in

$$I^2 R_{EMA} = F v \cos \phi_{MA} = 4.368 \times 10^{-3} \tag{4.2}$$

Since work equals power times time, multiplying both sides by the time taken to complete a cycle shows that work out per cycle equals work in per cycle

$$I^2 R_{EMA} T = F v \cos \phi_{MA} T = 9.57 \times 10^{-6} \tag{4.3}$$

This is in agreement with basic physics theory, where the work done per cycle is given as $W_{MACY} = \pi \omega R_{MEQ} A^2 = 9.56 \times 10^{-6}$. Then since work is just the transfer of energy, we see that energy is indeed conserved. We may account for the $I R_{EMA}$ drive voltage mathematically by noting that Kirchhoff's voltage law states that when tracing around a closed circuit, the algebraic sum of the potential changes encountered must be zero. By convention, a potential rise is positive and a potential drop is negative. Looking at the input circuit of Fig. 3.1 and noting that L_G and $R_G = $ zero, we find

$$E + \{[(I R_E) + (I R_{ECOR}) + (I R_{EMA})] + j(I X_{ET})\} = 0 \tag{4.4}$$

where $X_{ET} = \omega L_E$. This may be rearranged into a voltage equation where the sum of the magnitudes of all voltage drops is equal to the supply voltage

$$|E| = \sqrt{(I R_E + I R_{ECOR} + I R_{EMA})^2 + (I X_{ET})^2} = 1.41 \tag{4.5}$$

A relationship between R_{EMA} and back emf voltage was given by Thomson [8] and Morse [9]. It may be derived here by substituting the motor power rule $I E_B \cos \phi_{MA}$ for mechanical plus acoustic output power $F v \cos \phi_{MA}$ in Eq. (4.2), and solving for R_{EMA}

$$R_{EMA} = \frac{E_B}{I} \cos \phi_{MA} = 0.100 \tag{4.6}$$

The back emf voltage E_B will be discussed in detail in a later dedicated section on the subject. Now R_{EMA} may also be found as total resistance minus all other resistances

$$R_{EMA} = (Z_{ET} \cos \phi_E) - (R_E + R_{ECOR})) = 0.100 \tag{4.7}$$

Using the power-to-loss ratio gives

$$R_{EMA} = \frac{PL}{PL + 1} R_{ET} = 0.100 \tag{4.8}$$

4.3 Core Resistance

The next term to be considered is R_{ECOR}. This represents the resistive component of the motor core (mainly the pole piece) due to eddy current flow. Core resistance is generally modeled by a lossy inductor in the input impedance circuit. If the actual core resistance is found, this will allow the inductance to be defined in a simplistic manner. This technique will also allow calculation of true input power and true efficiency. At 456 Hz the power dissipated as measured with a Valhalla 2101 digital power analyzer is 0.2845 watt. The power going into the coil $I^2 R_E = 0.2483$ watt plus that creating mechanical and acoustic power $I^2 R_{EMA} = 4.366 \times 10^{-3}$ watt yields a sum of only 0.2527 watt. The difference of 0.0318 watt is due to the coil coupling to the core, in turn causing power dissipation in the core of $I^2 R_{ECOR} = 0.0318$ watt. To find the core resistance (considering source resistance to be zero), we note that power in must equal power out $P = (I^2 R_{EMA}) + [I^2 (R_E + R_{ECOR})]$, and solving for core resistance gives

$$R_{ECOR} = \frac{P - (I^2 R_E + I^2 R_{EMA})}{I^2} = 0.728 \tag{4.9}$$

The power into the core then is $I^2 R_{ECOR} = 0.0318$, in agreement with the above.
Core resistance is findable in several ways. The relationship to the power-to-loss ratio is

$$R_{ECOR} = \frac{Z_{ET} \cos \phi_E}{PL + 1} - R_E = 0.728 \tag{4.10}$$

where

$$\phi_E = arc \cos (P/EI) = arc \cos (R_{ET}/Z_{ET})$$

or alternatively,

$$R_{ECOR} = \frac{R_{EMA}}{PL} - R_E = 0.728 \qquad (4.11)$$

or with applied voltage,

$$R_{ECOR} = \frac{E}{I} \cos \phi_E - (R_E + R_{EMA}) = 0.728 \qquad (4.12)$$

This is just the real part of impedance minus the sum of all other resistances except core resistance. Rearranging terms yields

$$R_{ECOR} = \frac{E \cos \phi_E - [I(R_E + R_{EMA})]}{I} = 0.728 \qquad (4.13)$$

4.4 Blocked-Coil Measurement

In order to substantiate the above analysis with lab measurements, it was decided to find the resistance to eddy current flow in the core from measurements made with the coil blocked for no motion. To avoid destroying the reference driver, several units of the same model from the same production run were procured, and a unit closely matching the reference driver was selected. The coil was blocked with an epoxy based adhesive, with care taken that it remain in the equilibrium position where the measured inductance was the same as for the reference driver. Blocking the coil completely required epoxy both inside and outside of the coil, such that after the epoxy cured, no sound could be heard at any frequency with a stethoscope against the coil former. A comparison of the measured blocked-coil versus free-motion impedances is shown in the plot of Fig. 4.1.

Above mechanical resonance, the blocked-coil and free-coil impedances are almost identical, but vary slightly in three areas. The first range is between 200 and 600 Hz. Since blocked-coil impedance was slightly higher in this range, there was concern about the matching of the blocked-coil driver to the reference driver, so another driver of the same type was measured, first with motion and then with blocked-coil. The results were the same, with blocked-coil impedance slightly higher between 200 and 600 Hz. This odd effect has been noticed elsewhere [10]. It this case it appears to be due to mechanical resonance of the motor, basket, and mounting assembly (see Fig. 11.3, showing motor recoil velocity). Current with free versus blocked-coil conditions is shown in Fig. 4.2, where the current with blocked-coil is seen to be slightly less in the 200–600 Hz range, in accord with a marginally higher impedance. The other two differing points are around 900 and 1800 Hz where the current with free-coil dips slightly, in accordance with the impedance bumps in Fig. 4.1. The cause of this is attributed here to the first (going up in frequency) major

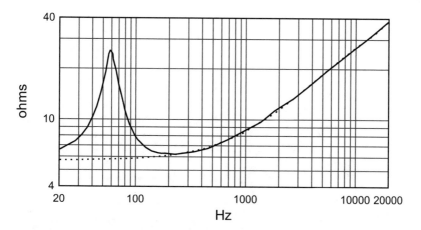

Fig. 4.1 Impedance versus frequency. Free (solid). Blocked-coil (dot)

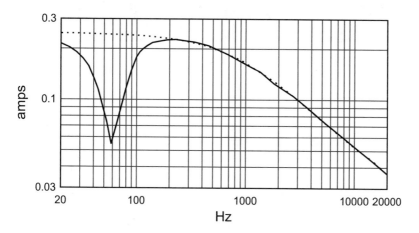

Fig. 4.2 Current versus frequency. Free (solid). Blocked-coil (dot)

stages of cone breakup. See Fig. 4.4 where the resistance due to the combined mechanical and acoustic loads R_{EMA} peaks at these frequencies.

The use of a non-magnetized unit similar to the reference driver was considered for blocked-coil measurements, but it was thought the absence of a magnetic field would affect current and power. To investigate this, a coil of 19 gauge wire was constructed to fit around the magnet of the non-reference blocked-coil driver. Applying DC voltage to the coil to affect the magnetic field strength of the motor did in fact affect current and power, and markedly so. With input voltage constant at 1.41 volts, a stronger magnetic field increases current and power, while decreasing impedance. Inductance was also affected, with a stronger field decreasing the inductance. In view of these inconsistencies, the use of a non-magnetized unit for measurements without motion should be approached with caution. Figure 4.3 provides a close-up of the blocked-coil test driver.

Fig. 4.3 Test driver with blocked coil

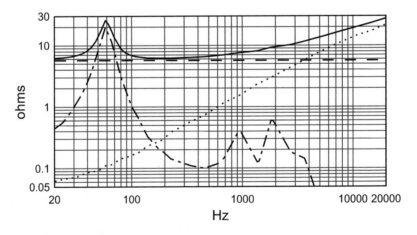

Fig. 4.4 Resistance versus frequency. R_{ET} (solid), R_E (dash), R_{ECOR} (dot), R_{EMA} (dash-dot)

With the coil blocked for no motion the measured current I_B at 456 Hz was 0.2067 amp. Free-coil current is 0.2090 amp, in accordance with the slightly higher imped-ance of the blocked-coil unit in Fig. 4.1. These values give the power dissipated in the coil of $I_B^2 R_E = 0.2429$ watt. The total power dissipated as measured with the power analyzer with the blocked-coil driver was 0.274 watt. The difference of 0.0311 watt is power into the core, and is shown in Table 4.1 for several different frequencies, along with the percentage of power going into the core.

Table 4.1 P_B is the total power dissipated with blocked-coil as measured. $P_{COILB} = I_B^2 R_E$ is power into the coil, and $P_{CORB} = P_B - P_{COILB}$ is power into the core

f	P_B	P_{COILB}	P_{CORB}	Core power in %
29	0.349	0.344	0.005	1.43
58.3	0.342	0.335	0.007	2.05
114	0.335	0.324	0.011	3.28
228	0.315	0.295	0.020	6.35
456	0.274	0.243	0.031	11.31
912	0.211	0.168	0.043	20.38
1820	0.143	0.094	0.049	34.27
3640	0.093	0.046	0.047	50.54
7280	0.063	0.022	0.041	65.08
14,560	0.044	0.011	0.033	75.00

Table 4.2 Core resistance versus frequency with blocked coil. Increase is relative to one octave below

f	R_{ECOR}	Increase/octave
29	0.068	
58.31	0.112	1.65
114	0.187	1.67
228	0.385	2.06
456	0.728	1.89
912	1.472	2.02
1820	2.962	2.01
3640	5.771	1.95
7280	10.705	1.85
14,560	17.022	1.59

Note that the mechanical plus acoustic power with the free-coil driver $v^2 R_{MEQ}$ when added to the blocked-coil and core power is about 2% less than the total dissipated power shown in Table 8.1. This is attributed to a slightly less than perfect match between the blocked-coil driver and the reference driver. Power dissipated in the core is only 2.05% at f_O but increases as frequency increases to 75% at 14,560 Hz. It peaks around 1800 Hz, then diminishes with increasing frequency. Note however, that the power into the core becomes greater than power dissipated in the coil above this point.

What we are seeking here is the actual core resistance, i.e. the component of resistance purely due to eddy current flow in the core. By means of measured power and current with a blocked-coil driver this is

$$R_{ECOR} = \frac{P_{CORB}}{I_B^2} = \frac{0.0311}{0.2067^2} = 0.728 \qquad (4.14)$$

in agreement with Eqs. (4.9) through (4.13).

The core resistance increases with frequency, in this case averaging an increase of a factor of 1.85 per octave between 29 and 14,560 Hz, and doubling per octave in the mid-range. This is shown in Table 4.2.

4.5   Source and Total Resistance

The final resistive component to be covered is R_G. This is the effective internal resistance of the source, which with most solid-state amplifiers is negligible, and is generally neglected. Such is the case here with the assumption that R_G = zero. Note we also consider the resistance of the connecting wire between the source and driver and its contact resistance as negligible and equal to zero. The total electrical resistance seen by the source at 456 Hz is thus

$$R_{ET} = R_E + R_{ECOR} + R_{EMA} + R_G = 6.513 \qquad (4.15)$$

in agreement with

$$R_{ET} = Z_{ET} \cos \phi_E = 6.513 \qquad (4.16)$$

Now the total resistance is given using the power-to-loss ratio as

$$R_{ET} = (PL + 1)(R_E + R_{ECOR}) = 6.513 \qquad (4.17)$$

The resistances are shown in Fig. 4.4. Note R_{EMA} peaks at 900 and 1800 Hz, coinciding with the impedance rises in Fig. 4.1. As noted, this appears to be due to the first major stages of cone breakup.

Looking at Eq. (4.16), since $Z_{ET} = E/I = 6.7464$ ohms and $\phi_E =$ arc cos $P/(EI) = 15.11°$, the input resistance may be found at any frequency with voltage, current, and power measurements.

The electrical phase angle ϕ_E is the phase angle between the AC voltage and current, and cos ϕ_E is known as the electrical power factor PF_E. The power factor is the part of impedance that is real and can absorb power, and accordingly the portion of applied volt-amperes EI that becomes real and usable power. Note however that all the usable power may not be directly useful in a given application. This is the case here, since the mechanical power is just a means to an end in order to obtain acoustic power. The electrical phase angle of the reference driver is plotted from 20 through 20,000 Hz in Fig. 4.5.

4.6   Reactance

The five elements of reactance are: (1, 2) $X_E + X_{ECOR}$, the coil and core reactances; (3, 4) X_{EMA}, the reactance due to the combined mechanical and acoustic loading of the driver; and (5) X_G, the source reactance. Components 1 and 2, $X_E + X_{ECOR}$, represent the reactances of the coil and core respectively. The blocked-coil inductance L_{EB} was measured at 456 Hz as $8.63 \times 10^{-4} H$, giving reactance of $\omega L_{EB} = 2.473$. The corresponding calculated reactance value is in close agreement:

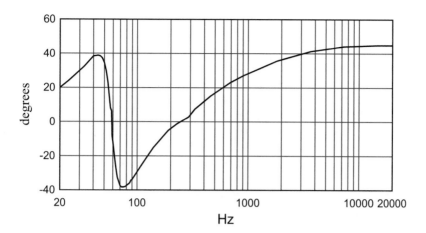

Fig. 4.5 Electrical phase angle

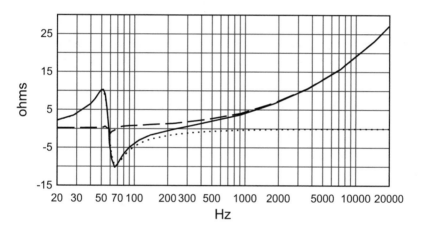

Fig. 4.6 X_{ET} (solid), X_{EMA} (dot), $X_E + X_{ECOR}$ (dash)

$$X_E + X_{ECOR} = (Z_{ET} - jR_{ET}) - X_{EMA} = 2.488 \qquad (4.18)$$

This is the total reactance, minus that due to motion. Note X_{EMA} in the above equation is negative. Reactances are shown in the plot of Fig. 4.6. The reason for the small blip in $X_E + X_{ECOR}$ at mechanical resonance is not known.

The core reactances (3, 4) are embodied in X_{EMA}. This is an equivalent reactance seen by the source due to the mechanical and acoustic (motional) reactances. X_{EMA} is given at 456 Hz as [11]

$$X_{EMA} = (Bl)^2 (X_{MEC} + X_A) / (Z_{MEC} + Z_A)^2 = -0.729 \qquad (4.19)$$

Reflection of this quantity to the electrical side of the equivalent circuit reverses the nature of the motional reactance [11–13], such that the positive mass reactance of

the load is seen by the source as a negative electrical reactance. Likewise the
negative compliance reactance of the load is seen by the source as positive reactance.
The load is mass-dominated and positive above mechanical resonance, while it is
compliance-dominated and negative below mechanical resonance. The coil and core
reactance $X_E + X_{ECOR}$ is positive at all frequencies. Above electrical resonance
(228 Hz), the positive coil and core reactance is greater than the negative (as seen
by the source) load reactance, making the net reactance X_{ET} positive (see Fig. 4.6).
Between electrical resonance and mechanical resonance the negative (as seen by the
source) load reactance is greater than the positive coil and core reactance, making the
net reactance X_{ET} negative. Below mechanical resonance, the load reactance is still
greater than the coil and core reactance, but the load reactance has now turned
positive (as seen by the source) so that the net reactance X_{ET} is again positive. Note
that at electrical resonance $f_E = 228$ Hz, the positive coil plus core reactance and the
negative (as seen by the source) load reactance are equal in magnitude, and being in
opposition to each other, they effectively cancel each other out. This leaves the net
reactance X_{ET} to be zero, such that voltage and current are in phase, impedance is
purely resistive ($Z_{ET} = R_{ET}$), the electrical power factor is unity, and EI is real
power. The same situation is true at mechanical resonance $f_O = 58.31$ Hz.

The motional reactance of the load as seen by the source may also be found as the
total reactance, minus that of the coil and core. At the 456 Hz reference frequency
this is

$$X_{EMA} = (Z_{ET} \sin \phi_E) - (X_E + X_{ECOR}) = -0.729 \qquad (4.20)$$

Component (5), X_G. This is the source reactance, and like source resistance is
considered to be zero for this analysis, as is connecting wire and connections. The
total reactance at 456 Hz is thus

$$X_{ET} = X_E + X_{ECOR} + X_{EMA} = 1.759 \qquad (4.21)$$

As noted, X_{EMA} is negative at 456 Hz. The above is in agreement with the basic
definition, where

$$X_{ET} = \frac{E}{I} \sin \phi_E = 1.759 \qquad (4.22)$$

Using the power-to-loss ratio gives

$$X_{ET} = Z_{ET} - j\left[(PL + 1)(R_E + R_{ECOR})\right] = 1.758 \qquad (4.23)$$

Knowing all the resistive elements of the impedance, and since reactance is of the
form $X = Z - jR$, then as a check, we find total reactance to be in agreement

$$X_{ET} = Z_{ET} - jR_{ET} = 1.759 \qquad (4.24)$$

4.7 Inductance

Inductance at 456 Hz is calculated as

$$L_E = \frac{X_{ET}}{\omega} = 6.14 \times 10^{-4} \tag{4.25}$$

This is the total inductance due to the reactance of the coil and core, plus the reactance of the mechanical and acoustic load as reflected to the electrical side. Using a General Radio 1650A impedance bridge, the measured inductance with motion was 6.12×10^{-4} for good agreement with the analytical values. The inductance is generally characterized as being lossy; however, we have already accounted for loss due to eddy current in the core with R_{ECOR}. Since $X_{ET} = (E/I) \sin \phi_E$, substituting this for X_{ET} in Eq. (4.25) allows inductance to be written as

$$L_E = \frac{E}{I\omega} \sin \phi_E = 6.14 \times 10^{-4} \tag{4.26}$$

Now since $\phi_E = arc \cos P/(E\,I)$, this allows calculation of L_E at any frequency in terms of voltage, current, and power. Note at the electrical and mechanical resonance frequencies, L_E in effect equals zero. Table 4.3 compares L_E per Eq. (4.26) with measured L_E values, where f, E, I and P were also measured. Regarding the measured L_E values, note the inductance meter employed for the measurements is slightly less accurate at the highest frequencies. Reactance X_{ET} is also shown in Table 4.3, where

Table 4.3 Comparison of L_E per Eq. (4.26) with measured L_E, where f, E, I, and P were also measured. Note reactance X_{ET} increases in the upper range as approximately $\sqrt{2}$ per octave

f	E	I	P	ϕ_E	L_E measured	L_E per Eq. (4.26)	$X_{ET} = \omega L_E$
20	1.41	0.2146	0.2850	19.63	0.0175	0.0176	2.212
28	1.41	0.1906	0.2370	28.13	0.0193	0.0198	3.483
40	1.41	0.1349	0.1490	38.43	0.0248	0.0258	6.484
58.3	1.41	0.0535	0.0755	0	0	0	0
110	1.41	0.1928	0.2460	−25.19	−0.00452	−0.00450	−3.110
140	1.41	0.2174	0.2960	−15.06	−0.00192	−0.00192	−1.689
228	1.41	0.2270	0.3200	0	0	0	0
456	1.41	0.2090	0.2845	15.11	0.000612	0.000614	1.759
912	1.41	0.1679	0.2125	26.15	0.000644	0.000646	3.702
1820	1.41	0.1240	0.1420	35.69	0.000570	0.000580	6.633
3640	1.41	0.0901	0.0940	42.28	0.000455	0.000460	10.521
7280	1.41	0.0623	0.0630	44.18	0.000346	0.000345	15.781
14,560	1.41	0.0431	0.0430	44.96	0.000265	0.000253	23.145
29,120	1.41	0.0300	0.0295	45.00	0.000199	0.000182	33.300

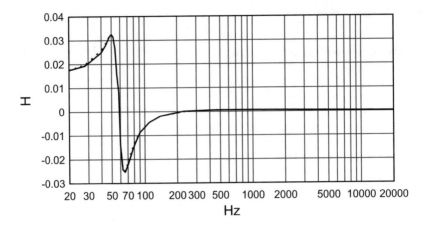

Fig. 4.7 Inductance. Measured (solid). Calculated (dot)

it can be seen in the upper range to reach a rate of increase of approximately $\sqrt{2}$ per octave.

Since reactance $X = Z - jR$, the inductance may also be expressed as

$$L_E = (Z_{ET} - jR_{ET})/\omega = 6.14 \times 10^{-4} \tag{4.27}$$

Otherwise with voltage, current, and power

$$L_E = \left[(E/I)\sqrt{1 - (P/EI)^2} \right]/\omega = 6.14 \times 10^{-4} \tag{4.28}$$

or as

$$L_E = \{Z_{ET} - j\left[(PL + 1)(R_E + R_{ECOR})\right]\}/\omega = 6.14 \times 10^{-4} \tag{4.29}$$

L_E per Eq. (4.26) versus measured L_E is also shown in graphical form in Fig. 4.7, and with an expanded scale in Fig. 4.8.

The impedance bridge would not register any inductance between f_O and f_E. This is due to the fact that between the mechanical and electrical resonance points, the positive mechanical plus acoustic reactance is reflected to the electrical side as negative (capacitive), and since that term is larger than the positive coil reactance, the source sees the net reactance as capacitive. Setting the meter to the capacitance mode gave a reading of 674 µF at 140 Hz, which translates to an equivalent negative inductance of

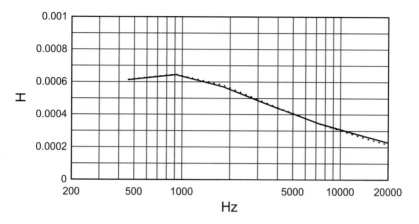

Fig. 4.8 Inductance versus frequency (expanded). Measured (solid). Calculated (dot)

$$L_{E(EQ)} = \frac{1}{\omega^2 C} = -1.92 \times 10^{-3} \tag{4.30}$$

this being obtained by substitution of $X = 1/(\omega C)$ into $L = X/\omega$, and in close agreement with the calculated inductance of

$$L_E = \frac{E}{I\omega} \sin \phi_E = -1.93 \times 10^{-3} \tag{4.26}$$

where at 140 Hz, the current $I = 0.217$ and $\phi_E = -15.06$.

4.8   Impedance Expressions

We now have the tools to write an expression for the total input impedance that includes all the parameters involved:

$$|Z_{ET}| = \sqrt{(R_G + R_E + R_{ECOR} + R_{EMA})^2 + (X_G + X_E + X_{ECOR} + X_{EMA})^2} = 6.7424. \tag{4.31}$$

which can be combined as the basic expression

$$Z_{ET} = R_{ET} + jX_{ET} = 6.7424 \tag{4.32}$$

The actual measured impedance was $Z_{ET} = E/I = 6.7464$. For comparison, the analogous circuit of Fig. 3.1 gives

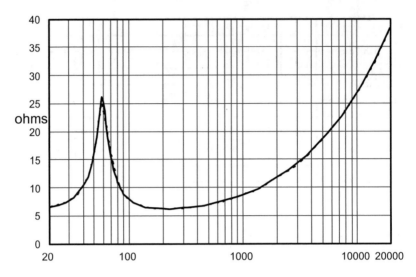

Fig. 4.9 Impedance. Measured as E/I (solid). Calculated per Eq. (4.33) (dot)

$$|Z_{ET}| = \sqrt{(R_G + R_E + R_{ECOR} + R_{EMA})^2 + [\omega\,(L_E + L_G)]^2} = 6.7465 \qquad (4.33)$$

or simplifying

$$Z_{ET} = R_{ET} + j[\omega\,(L_E + L_G)] = 6.7465 \qquad (4.34)$$

where $L_E = [E/(I\,\omega)]\,\sin\,\phi_E$. The power-to-loss ratio yields

$$Z_{ET} = (R_E + R_{ECOR})\,(PL + 1)/\cos\,\phi_E = 6.7464 \qquad (4.35)$$

The impedance per Eq. (4.33) is compared with measured impedance $Z_{ET} = E/I$ in Fig. 4.9. This is also shown for several frequencies in Table 4.4, where R_{ET} is included for reference. Significantly, there is almost an exact agreement between the values predicted by Eq. (4.33) and the measured values.

4.9 Voltage and Power Expressions

From the foregoing analysis, a voltage equation may be written as the sum of the magnitudes of all voltage drops across the resistive and reactive elements:

Table 4.4 Comparison of measured $Z_{ET} = E/I$ with Eq. (4.33), where $L_E = [E/(I\omega)]\sin\phi_E$

f	R_{ET}	$Z_{ET} = E/I$	Z_{ET} per Eq. (4.33)
20	6.19	6.57	6.57
28	6.57	7.40	7.41
40	8.29	10.45	10.42
58.3	25.71	26.16	25.71
110	6.65	7.31	7.34
140	6.26	6.49	6.48
228	6.21	6.21	6.21
456	6.51	6.75	6.75
912	7.54	8.40	8.40
1820	9.24	11.37	11.33
3640	11.60	15.65	15.58
7280	16.38	22.63	22.56
14,560	22.99	32.71	32.41
20,000	27.32	38.58	38.56

$$E = [(IR_G) + (IR_E) + (IR_{ECOR}) + (I_{REMA})] + j\{[I(X_E + X_{ECOR})] + (IX_{EMA})\} = 1.41 \tag{4.36}$$

or simplifying,

$$E = (IR_{ET}) + j(IX_{ET}) = 1.41 \tag{4.37}$$

and as related to back emf voltage E_B in magnitudes

$$E = \{[I(R_G + R_E + R_{ECOR})] + (Blv\ \cos\ \phi_{MA})\} + j(IX_{ET}) = 1.41 \tag{4.38}$$

or by using the power-to-loss ratio,

$$E = (PL + 1/PL)\ (IR_{EMA})/\cos\ \phi_E = 1.41 \tag{4.39}$$

where the term IR_{EMA} represents the drive voltage E_{DR}, i.e. the portion of the source voltage driving the mechanical and acoustic load. As related to the input voltage, the real or true power delivered by the source is

$$P = EI\ \cos\ \phi_E = 0.2845 \tag{4.40}$$

which is in close agreement with true power as measured with the power analyzer.

Chapter 5
Efficiency and the Mechanical Response Function

5.1 Efficiency at the Working Frequency

Loudspeaker efficiency is generally stated with reference to the middle of the piston-band. Let's look first however, at the frequency we are working with of 456 Hz. Since power divides between two resistances according to the voltage divider rule, and noting R_{MEQ} includes both mechanical and acoustic resistance, the acoustic power output at 456 Hz on one side of the cone is:

$$P_A = (P_{MEC} + P_A) \frac{R_{MA}}{R_{MEQ}} = 1.112 \times 10^{-3} \tag{5.1}$$

which is in agreement with Beranek [14]

$$P_A = v^2 R_{MA} = 1.112 \times 10^{-3} \tag{5.2}$$

The true input power from the source is given as

$$P = \frac{E^2}{R_{ET}} (\cos \phi_E)^2 = I^2 R_{ET} = EI \cos \phi_E = 0.2845 \tag{5.3}$$

again in agreement with true power as measured. True efficiency then, is the ratio of acoustic power radiated from the front side of the cone to the electrical input power, such that at 456 Hz

$$\eta = \frac{P_A}{P} = 0.00391 = 0.391\% \tag{5.4}$$

Figure 5.1 shows a plot of the reference driver efficiency from 20 through 912 Hz. Note that maximum efficiency occurs slightly above the mechanical resonance frequency.

Replacing the coil resistance R_E with total resistance R_{ET} in Small's passband efficiency expression [15], and including the mechanical response function $MRF = G(j\omega)(m)$ gives an efficiency value of

$$\eta = \frac{\rho_0}{2\pi c} \frac{(Bl)^2}{R_{ET}} \frac{S_D^2}{M^2} |G(j\omega)(m)|^2 = 0.00391 = 0.391\% \qquad (5.5)$$

The overall transduction efficiency per Eq. (5.5) compared to P_A/P along with related data is shown in Table 5.1, where it may be seen that the mechanical response function allows Eq. (5.5) to be valid at all frequencies.

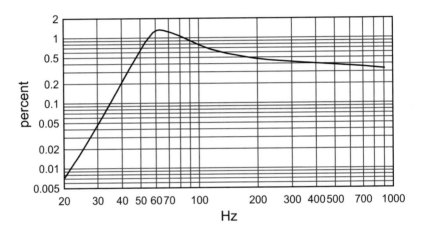

Fig. 5.1 Efficiency versus frequency

Table 5.1 Efficiency per Eq. (5.5) compared to the ratio of power out to power in $P_A = v^2 R_{MA}$, $R_{MA} = (\rho_0/2\pi c)\omega^2 S_D^2$, $P = I^2 R_{ET}$. Constants needed in Eq. (5.5) are $S_D = 0.0238$, $Bl = 7.31$, $M = 0.02572$. The remaining values needed in Eq. (5.5) are shown in the table for the reader's convenience

f	$\eta = P_A/P$	η per Eq. (5.5)	v	I	R_{ET}	$G(j\omega)(m) = \frac{Ma}{F}$	a	F
40	0.210	0.210	0.1279	0.1353	8.292	0.836	32.141	0.9890
58.3	1.210	1.208	0.1502	0.0548	25.706	3.531	55.041	0.4009
140	0.568	0.567	0.0839	0.2174	6.259	1.194	73.770	1.5892
228	0.453	0.453	0.0479	0.2270	6.211	1.063	68.607	1.6594
456	0.391	0.391	0.0210	0.2090	6.513	1.011	60.056	1.5278

5.2   The Mechanical Response Function

The mechanical response function $MRF = G(j\omega)(m)$ as used here accounts for force differing from mass times acceleration. This discrepancy is due to mechanical resonance affecting response at its frequency of occurrence f_o. The effect extends above and below f_o, but with diminishing influence. The mechanical response function may be viewed as the ratio of response in the region of mechanical resonance to that at infinity. The reason Newton's second law of motion $F = Ma$ is not directly applicable around mechanical resonance is that mass requires less force for a given acceleration. This is due to the increased mechanical efficiency of the system at and around mechanical resonance. Let's look at what happens at mechanical resonance. At an end point of cone travel, motion has stopped for an instant, but the potential energy in the stretched suspensions forces the mass to move toward the center of travel (equilibrium position). When the mass reaches the equilibrium position the suspensions are relaxed (no longer stretched), and now possess no stored potential energy. However the cone velocity is now at a maximum, such that the kinetic energy in the moving mass causes overshoot, and begins stretching the suspensions in the opposite direction. This continues until the kinetic energy in the mass is expended at the other end-point of travel. At this point motion stops, and the stored potential energy in the stretched suspensions again forces the mass to move, this time in the reverse direction. In other words, at resonance the energy accelerating the mass during one quarter cycle comes from the stretched suspensions, propelling the cone to the equilibrium rest position where the suspensions are un-stretched. Then during the next quarter cycle, that energy has been converted to kinetic energy in the moving mass that propels the cone onward. This begins re-stretching the suspensions in the opposite direction until the kinetic energy is expended, motion stops, and again all the energy is in the stretched suspensions. The process then repeats itself in the reverse direction for the next two quarter cycles, then starts over again as in the first direction, and so on. Therefore at mechanical resonance and once in motion, no input power (energy) is required to move the mass, and the only external energy input needed to maintain steady state motion is that required to overcome system friction (the loss resistances). We could say that mass effectively gets a free ride around resonance.

Regarding the deviation of force from mass times acceleration, the mechanical response function may be stated as the ratio of inertial force Ma to the electromagnetic driving force F

$$G(j\omega)(m) = \frac{Ma}{F} = 1.011 \tag{5.6}$$

As related to suspension stiffness we have

$$G(j\omega)(m) = r^2 \frac{kA}{F_{(MAX)}} = 1.011 \tag{5.7}$$

where $r = \omega/\omega_O$, the ratio of operating frequency ($2\pi f$) to the resonance value ($2\pi f_O$).

The mechanical response function may also be found from the ratio of mass reactance to equivalent mechanical plus acoustic resistances

$$G(j\omega)(m) = \frac{\omega M}{R_{MEQ}} \cos \phi_{MA} = 1.011 \tag{5.8}$$

Note at f_O, $\cos \phi_{MA} = 1$, and the magnitude of the mechanical response function is thus equal to Q_M, which then asymptotically approaches unity at higher frequencies. With volume acceleration we get:

$$G(j\omega)(m) = \frac{M}{F S_D} V_a = 1.011 \tag{5.9}$$

Perhaps the most intuitive expression is the ratio of mass reactance to impedance

$$G(j\omega)(m) = \frac{\omega M}{(Z_{MEC} + Z_A)} = 1.011 \tag{5.10}$$

From the physics of mechanical vibrations, the following expression may be found in several variations:

$$G(j\omega)(m) = \frac{r^2}{\sqrt{(1 - r^2)^2 + (r/Q_M)^2}} = 1.016 \tag{5.11}$$

In practice, it is generally observed that the mechanical Q function, Q_M, is highly sensitive to even minute temperature and humidity variations. This makes it difficult to measure with absolute precision, even in lab environments. Therefore using measured values of Q_M are much more likely to result in variations in parametric calculations. The overall mechanical response function for the reference driver using Eq. (5.10) is shown in Fig. 5.2.

5.3   Relation of Above Parameters and Power

Now back to considerations of efficiency, adding R_{ECOR} to R_E in the equation for efficiency from Kinsler et al. [16] gives the true efficiency of the reference driver at 456 Hz of

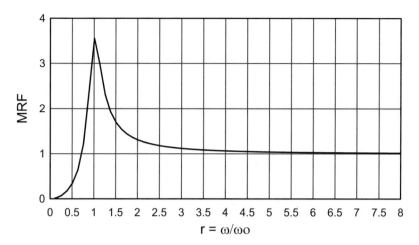

Fig. 5.2 Mechanical response function

$$\eta = \frac{R_{MA}}{\left[(R_E + R_{ECOR})(Bl)^2/Z_{EMA}^2\right] + R_{MEQ}} = 0.00391 = 0.391\% \qquad (5.12)$$

where R_{MEQ} was noted by Kinsler as $R_M + R_{MA}$. The following values at 456 Hz are noted here for convenience,

$R_{MA} = 2.5312$
$(R_E + R_{ECOR}) = 6.4131$
$Z_{EMA} = (Bl)^2/(Z_{MEC} + Z_A) = 0.73315$
$R_{MEQ} = (Z_{MEC} + Z_A) \cos \phi_{MA} = 9.9234$
$Z_{MEC} + Z_A = Bll/v = 72.8859$
$I = 0.209$
$v = a/\omega = Va/\omega S_D = 0.02096$

where velocity was verified by laboratory measurements with a Polytec PDV-100 laser vibrometer.

Efficiency may be found with volume acceleration as

$$\eta = \frac{po}{2\pi c} \frac{Va^2}{P} = 0.00391 = 0.391\% \qquad (5.13)$$

or alternatively by using the power-to-loss ratio,

$$\eta = \frac{PL}{PL + 1} \frac{R_{MA}}{R_{MEQ}} = 0.00391 = 0.391\% \qquad (5.14)$$

or with mechanical and acoustic power

$$\eta = \frac{P_{MEC} + P_A}{P} \frac{R_{MA}}{R_{MEQ}} = 0.00391 = 0.391\% \tag{5.15}$$

Another confirmation may be derived with work done per cycle on the mechanical and acoustic load

$$\eta = \frac{W_{EMACY}f}{P} \frac{R_{MA}}{R_{MEQ}} = 0.00391 = 0.391\% \tag{5.16}$$

or as related to kinetic energy,

$$\eta = KE_{(MAX)} \frac{R_{MA}}{MP} = 0.00391 = 0.391\% \tag{5.17}$$

Now let's look at efficiency in the middle of the piston-band, specifically at the low point in impedance $f_{ZMIN} = f_E$ located in frequency above mechanical resonance. This may be defined as the frequency of electrical resonance. For the reference driver this is 228 Hz, where the acoustic power output on one side of the cone is

$$P_A = (P_{MEC} + P_A) \frac{R_{MA}}{R_{MEQ}} = 1.45 \times 10^{-3} \tag{5.1}$$

where

$P_{MEC} + P_A = 7.28 \times 10^{-3}$
$R_{MA} = 0.6328$
$R_{MEQ} = 3.1738$

otherwise given by Beranek as [14]

$$P_A = v^2 R_{MA} = 1.45 \times 10^{-3} \tag{5.2}$$

where $v = 0.04789$, as measured with the vibrometer.

Looking into the process that determines input power, recall that above mechanical resonance the load is dominated by positive mass reactance, which is reflected into the electrical circuit reversed in nature as a negative reactance. This negative reactance X_{EMA} is equal in magnitude to, but opposing the positive inductive reactance $X_E + X_{ECOR}$ of the coil and core at f_{ZMIN}, such that electrical reactance is in effect canceled out (see Fig. 4.6). This is the frequency of electrical resonance f_E where voltage and current are in phase, i.e. the electrical phase angle is zero. Since this occurs in the middle of the piston-band, it is a logical point to specify efficiency. Now since the phase angle is zero ($\cos \phi_E = 1$), $\cos \phi_E$ drops out of the power equations of Eq. (5.3) as given earlier. Accordingly, with efficiency at $f_{ZMIN} = f_E$ considered the reference efficiency, the true reference input power utilized for calculating true reference efficiency is (note the following in this section uses magnitudes at f_E of 228 Hz as given in Table 5.1):

$$P = \frac{E^2}{R_{ET}} = I^2 R_{ET} = EI = 0.320 \tag{5.18}$$

which is in agreement with power as measured with the power analyzer, and where $I = 0.227$, and R_{ET} is

$$R_{ET} = \frac{E}{I} = 6.211 \tag{5.19}$$

True reference efficiency is the ratio of acoustic power radiated from the front side of the cone to true input power at f_{ZMIN}, such that

$$\eta_O = \frac{P_A}{P} = 0.00453 = 0.453\% \tag{5.20}$$

or in terms of basic driver parameters, analogous to Eq. (5.5)

$$\eta_O = \frac{\rho_O}{2\pi c} \frac{(Bl)^2}{R_{ET}} \frac{S_D{}^2}{M^2} |G(j\omega)(m)|^2 = 0.00453 = 0.453\% \tag{5.5}$$

where the mechanical response function at 228 Hz is now

$$G(j\omega)(m) = \frac{\omega M}{R_{MEQ}} \cos \phi_{MA} = \frac{Ma}{F} = 1.0634 \tag{5.8}$$

As seen, Table 5.1 compares efficiency with Eq. (5.5) to P_A/P and as noted, shows that the mechanical response function makes Eq. (5.5) valid at all frequencies.

5.4   Correlation with Kinsler, Kloss, and Small

Adding R_{ECOR} to R_E in the equation for efficiency by Kinsler et al. [16] gives a reference efficiency of

$$\eta_O = \frac{R_{MA}}{\left[(R_E + R_{ECOR})\,(Bl)^2/Z_{EMA}{}^2 \right] + R_{MEQ}} = 0.00453 = 0.453\% \tag{5.21}$$

where at 228 Hz

$R_{MA} = 0.6328$
$R_E + R_{ECOR} = 6.0702$
$Z_{EMA} = (Bl)^2/(Z_{MEC} + Z_A) = 1.5422$
$R_{MEQ} = (Z_{MEC} + Z_A) \cos \phi_{MA} = 3.1735$
$Z_{MEC} + Z_A = Bll/v = 34.649$

According to Kloss [17], in the mid 1950s Dr. J. Anton Hofmann introduced the term "damping frequency", which allowed the formulation of Hofmann's Iron Law. Kloss gave this in a quantitative expression for efficiency as $\eta_O \approx f_R^2 f_D V$, where F_R = mechanical resonant frequency, f_D = damping frequency, and V = box volume. In the notation used here and in the same order, this is $\eta_O \approx f_O^2 f_3 V_B$. By adding the proper constants, accounting for total system compliance by using V_{AT} instead of V_B (per Small below [18]), while also accounting for a given Q_T, and using the measured value of f_3, the expression gives:

$$\eta_O = \frac{4\pi^2}{c^3} \frac{f_O^2 f_3 V_{AT}}{\sqrt{Q_T}} = 0.00445 = 0.445\% \tag{5.22}$$

where measured $f_3 = 52.6$ Hz, and $V_{AT} = 0.02304$. This is off from P_A/P by only 1.8%.

A useful method of circumventing this trade-off between efficiency and bass extension in a given cabinet volume is given in Chap. 16.

Small [18] gave the expression for reference efficiency as

$$\eta_O = \frac{4\pi^2}{c^3} \frac{f_O^3 V_{AT}}{Q_E} = 0.00430 = 0.430\% \tag{5.23}$$

where in his notation $f_O = f_C$ and $Q_E = Q_{EC}$. This is off from P_A/P by 5.1%. As noted previously, measurements around mechanical resonance f_O are significantly more prone to variation due to the driver's extreme sensitivity to temperature and humidity at this frequency.

Finally, from basic motor theory, we find efficiency must be $\eta = 1 - [(P - P_A)/P] \times 100 = 0.453\%$.

5.5 Efficiency Expressions and Loss

True reference efficiency may be found by using volume acceleration as

$$\eta_O = \frac{\rho_O}{2\pi c} \frac{Va^2}{P} = 0.00453 = 0.453\% \tag{5.24}$$

using the power-to-loss ratio yields

$$\eta_O = \frac{PL}{PL+1} \frac{R_{MA}}{R_{MEQ}} = 0.00453 = 0.453\% \tag{5.14}$$

or alternately with mechanical and acoustic power

$$\eta_O = \frac{P_{MEC} + P_A}{P} \frac{R_{MA}}{R_{MEQ}} = 0.00454 = 0.454\% \tag{5.25}$$

Also confirmation with the work done per cycle on the mechanical and acoustic load gives

$$\eta_O = \frac{W_{EMACY} f}{P} \frac{R_{MA}}{R_{MEQ}} = 0.00453 = 0.453\% \tag{5.26}$$

Finally, the relation to kinetic energy is

$$\eta_O = KE_{(MAX)} \frac{R_{MA}}{MP} = 0.00453 = 0.453\% \tag{5.27}$$

The power conversion process gives an acoustic output yield in the piston-band of only one part out of 221 parts of the electrical input energy started with, and the loss of power in the piston-band is

$$Loss = 1 - \eta_O = 0.9955 = 99.55\% \tag{5.28}$$

This equation makes clear the startling conversion inefficiency of the typical moving coil direct-radiator loudspeaker. We can look at its acoustic output as just air that's pushed aside, then pulled back by a moving mass, with the mass itself being moved back and forth by a motor producing mostly heat. The process is much like shaking the trunk of a small tree just to hear the rustling of the leaves.

As noted earlier, a novel method of circumventing the trade-off between efficiency and bass extension in a given cabinet volume is presented later in Chap. 16.

Chapter 6
Back EMF

6.1 Faraday's and Lenz's Laws

Faraday's law states that movement of a conductor in a magnetic field will cause an induced voltage in the conductor. Therefore during motion the voice coil behaves as the armature of an electrical generator, producing a voltage of its own. This induced voltage is known as back emf, and also as counter emf or speed voltage. According to Lenz's law and as related to voltage, the back emf voltage acts to oppose the applied voltage.

6.2 Transfer of Energy

There are several ways to view the function of back emf. Fitzgerald [19] stated that back emf is responsible for energy transferred in electromechanical energy conversion systems. Accordingly, let's look first at energy transfer from the perspective that power is dissipated into a resistance in order to move the mechanical and acoustic load. This resistance may be found by first converting the mechanical and acoustic impedance to the electrical side [20].

$$Z_{EMA} = \frac{(Bl)^2}{Z_{MEC} + Z_A} = 0.7331 \tag{3.3}$$

Then since force is not in phase with velocity, the source sees the real part of this as

W. H. Watkins, *Loudspeaker Physics and Forced Vibration*,
https://doi.org/10.1007/978-3-030-91634-3_6

$$R_{EMA} = \frac{(Bl)^2}{Z_{MEC} + Z_A} \cos \phi_{MA} = 0.100 \tag{3.4}$$

Thus the resistance due to mechanical and acoustic impedance is reflected into the electrical circuit and seen by the source as an effective electrical resistance. This resistance is given with the power-to-loss ratio as

$$R_{EMA} = PL \left(R_E + R_{ECOR}\right) = 0.100 \tag{6.1}$$

Now let's look at the portion of applied voltage associated with driving the combined mechanical and acoustic loads. Per Fig. 3.1 and noting R_G and L_G are assumed to be zero, the load resistance R_{EMA}, coil resistance R_E, core resistance R_{ECOR}, and reactance ωL_E form a voltage divider such that the voltage drop across R_{EMA} is:

$$I R_{EMA} = E \; \frac{R_{EMA}}{\sqrt{\left(R_{EMA} + R_E + R_{ECOR}\right)^2 + \left(\omega L_E\right)^2}} = 0.0209 \tag{6.2}$$

It follows that electrical power driving the mechanical and acoustic load is $I^2 R_{EMA}$, such that power in and power out are equal:

$$I^2 R_{EMA} = v^2 R_{MEQ} = 4.368 \times 10^{-3} \tag{6.3}$$

Then since power is energy transferred per second, energy out is equal to energy in, and energy is conserved.

6.3   Degree of Current Flow

Now the question arises as to what sets the degree of current flow in Eq. (6.3), and this is where back emf enters the picture. Let's look at the process involved. At an end point of travel where the coil and cone have stopped for a brief instant to turn around, there is no motion-dependent back emf and current is maximum, being limited only by the impedance of the coil and core. The current produces an electromagnetic force and acceleration, which starts the generation of back emf. Now this back emf opposes the drive voltage, causing current to decrease. Then as velocity and back emf increase this decreases current until just enough flows to overcome the losses in the motor $R_E + R_{ECOR}$, the resistance due to the mechanical and acoustic load R_{EMA}, and the system reactance X_{ET}:

$$I = \frac{E}{(R_E + R_{ECOR} + R_{EMA}) + jX_{ET}} = 0.209 \qquad (6.4)$$

in agreement with current as measured. From above we know that R_{EMA} is the resistance due to the mechanical and acoustic load as seen by the source. For a given load then, we may write:

$$I = \frac{E_B}{R_{EMA}} \cos \phi_{MA} = 0.209 \qquad (6.5)$$

where $E_B = Blv$. So we see that for a given mechanical and acoustic load, back emf controls the degree of current flow. Perhaps a more graphic example is shown with mechanical plus acoustic impedance:

$$I = \frac{E_B}{(Bl)^2} (Z_{MEC} + Z_A) = 0.209 \qquad (6.6)$$

Then with $(Bl)^2$ constant and $Z_{MEC} + Z_A$ constant for a given frequency, again we see that back emf controls the degree of current flow, and thereby the amount of power driving the mechanical and acoustic load.

6.4 Newton's Laws

It is interesting that we can link back emf to Newton's laws. From acoustics theory, we know the magnitude of the mechanical plus acoustic load force is $F_L = v$ $(Z_{MEC} + Z_A)$. Then since $v = E_B/Bl$, we may again bring back emf into the picture:

$$F_L = \frac{E_B}{Bl} (Z_{MEC} + Z_A) = 1.527 \qquad (6.7)$$

Noting $(Z_{MEC} + Z_A) = \omega M/MRF$, where MRF is the mechanical response function $G(j\omega)\,(m)$

$$F_L = \frac{E_B}{Bl} \frac{\omega M}{MRF} = 1.528 \qquad (6.8)$$

As noted, $E_B/Bl = v$, so $F_L = v\,\omega\,M/MRF$, but we recognize $v\,\omega$ as acceleration a, so

$$F_L = \frac{M a}{MRF} = 1.528 \qquad (6.9)$$

which is Newton's second law of motion with *MRF* allowing for mechanical resonance f_O. Then by Newton's first law for steady-state motion (in this case, no change in velocity from cycle to cycle), we know that the net force acting on the system must be zero. Since the driving force BlI is positive and the load force is actually negative, this is shown as:

$$F_{NET} = BlI + \frac{Ma}{MRF} = 0 \tag{6.10}$$

So we see that for a given applied voltage, forces are balanced as forces are wont to do, and the load moves along at steady-state.

6.5   Conservation of Energy

Looking at back emf from a purely voltage point of view, and starting from standstill at an end point of travel, velocity increases (which increases back emf and decreases current) until the voltage drop across R_{EMA}, plus the voltage drop across motor resistance, plus the voltage overcoming system reactance is equal in magnitude to the supply voltage:

$$\begin{aligned} E &= \{[I(R_E + R_{ECOR})] + (IR_{EMA})\} + j(IX_{ET}) \\ &= 1.41 \end{aligned} \tag{6.11}$$

This is just an example of energy conservation in terms of Kirchoff's voltage rule, where the sum of the voltage drops across series load elements is equal in magnitude to the applied voltage. The first term on the right in Eq. (6.11) is the voltage creating heat in the coil and core, the second term is the voltage driving the mechanical and acoustic load (equivalent in magnitude to the portion of back emf in phase with current $E_B \cos \phi_{MA}$), and the last term is the voltage overcoming system reactance.

Table 6.1 Comparison of the magnitudes of drive voltage IR_{EMA} and the component of back emf in phase with current $E_B \cos \phi_{MA}$.

f	IR_{EMA}	$E_B \cos \phi_{MA}$
40	0.3411	0.3410
58.3	1.0982	1.0982
140	0.0714	0.0712
228	0.0321	0.0321
456	0.0209	0.0209
912	0.0640	0.0641
1820	0.0745	0.0745
3640	0.0126	0.0126
7280	0.00274	0.00274
14,560	0.00075	0.00075

As Table 6.1 shows, the $I R_{EMA}$ drive voltage and the effective part of back emf are equal in magnitude.

Accordingly, and from a voltage point of view, what we are calling R_{EMA} in Fig. 3.1 may be replaced with $E_B \cos \phi_{MA} = Blv \cos \phi_{MA}$. This is the case in a circuit given by Leach [21]. Then with back emf representing the mechanical and acoustic load in Fig. 3.1 as $Blv = [(Bl)^2/(Z_{MEC} + Z_A)] I$, the electrical power driven into the load is:

$$P_{EMA} = I Blv \cos \phi_{MA} = 4.366 \times 10^{-3}, \tag{6.12}$$

which we recognize from books on motors as the de-facto expression for motor mechanical power $I E_B \cos \phi_{MA}$.

6.6 Back emf and Power Transfer

In the above respect and with regard to electrical motors in general, there is a concept sometimes stated that says electrical power working against or driven into back emf is the power that creates mechanical power [22–24]. This would, of course, include power going into the air resistance encountered by the load. Power driven into back emf would be power into the resistance presented to the source by back emf. That resistance is:

$$R_{EMA} = \frac{E_B}{I} \cos \phi_{MA} = 0.100 \tag{4.6}$$

giving power into the mechanical and acoustic load of

$$P_{EMA} = I^2 R_{EMA} = 4.368 \times 10^{-3}, \tag{6.13}$$

in agreement with Eq. (6.12). With regard to the electromagnetic driving force, back emf subtracts from the applied voltage such that:

$$F = Bl (E \cos \phi_E - E_B \cos \phi_{MA})/R_E + R_{ECOR} = 1.527, \tag{6.14}$$

in agreement with the earlier result $F = I E_B/v = 1.527$.

It is interesting to look at the electromagnetic driving force from a magnetic point of view, and assuming that back emf could produce current and a magnetic field of its own. Since back emf opposes the applied voltage and according to Lenz's law, any magnetic field its current would create would oppose the magnetic field created by current due to the applied voltage. It would go something like this: starting from standstill at an end point of travel, the applied voltage forces current through the coil, creating a magnetic field that works against the magnetic field of the motor. This creates an electromagnetic driving force that produces motion, which in turn starts the generation of back emf. Now the back emf would also create a magnetic field, but in opposition to that due to the applied voltage. Then as velocity increased, back emf would also increase until the magnetic field due to the back emf was equal in intensity to that due to the applied voltage. At this point the two opposing magnetic fields would be equal, leaving no more net driving force. Acceleration would thus stop, and both velocity and back emf would be at their maximum values:

$$v_{peak} = \frac{E_{B\ peak}}{Bl} = \frac{E_{B\ peak}}{\sqrt{\left(R_{MS} + 2R_{MA(\ f_o)}\right) R_{EMA(\ f_o)}}} = 0.02964 \qquad (6.15)$$

Concerning the effective part of back emf, back emf is in phase with velocity and the magnetizing drive current is in phase with force, but velocity and force are not always in phase, therefore neither is back emf and current. This being so, the effective portion of back emf is $E_B \cos \phi_{MA}$.

The generated back emf is:

$$E_B = \frac{E \cos \phi_E - [I (R_E + R_{ECOR})]}{\cos \phi_{MA}} = 0.1532 \qquad (6.16)$$

and the back emf as related to the power-to-loss ratio is:

$$E_B = E \frac{PL \cos \phi_E}{(PL + 1) \cos \phi_{MA}} = 0.1532 \qquad (6.17)$$

So what is the bottom line on the function of back emf? Since the objective is to transfer power from the electrical domain into mechanical and acoustic power, we note that electrical input power is simply:

$$P = EI \cos \phi_E = 0.2845 \ \text{watt} \qquad (4.40)$$

Then in keeping with conservation of energy, output power is:

$$P_{OUT} = \left[I^2 (R_E + R_{ECOR}) \right] + [(I E_{\mathbf{B}}) \cos \phi_{MA}] = 0.2845 \ \text{watt} \qquad (6.18)$$

So for a given applied voltage and load, back emf controls the degree of current flow and power transfer such that power out is equal to power in.

As noted earlier, the basic expression for back emf is $E_B = Blv$, and back emf is shown versus the drive frequency in Fig. 6.1.

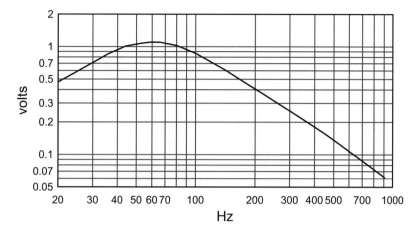

Fig. 6.1 Back emf versus frequency

Chapter 7
Mechanical and Acoustic Load Impedances

7.1 Mechanical Plus Acoustic Impedance

The combined mechanical impedance and acoustic impedance $Z_{MEC} + Z_A$ represents the opposition the source must overcome to produce motion. With reference to the elements on the mechanical and acoustic side, at 456 Hz this is:

$$|Z_{MEC} + Z_A| = \sqrt{(R_{MS} + 2R_{MA})^2 + (\omega M - 1/\omega C_{MT})^2} = 72.886 \qquad (7.1)$$

Note that the mechanical and acoustic masses are lumped together as M. Since the driver was not measured in a vacuum, the air load was moved along with the cone and coil, its mass is in effect being added to that of the cone and coil. One could say the air load was stuck to the cone, as it were. Likewise, the air load compliance is lumped with the suspension compliance in the term C_{MT}. This of course, is the condition of normal speaker operation, where the movement of the cone, coil, and air load always occurs simultaneously. The power-to-loss ratio gives the combined mechanical and acoustic impedances as:

$$Z_{MEC} + Z_A = \frac{PL}{PL + 1} \frac{P}{v^2 \cos \phi_{MA}} = 72.877 \qquad (7.2)$$

The relation to the equivalent mass reactance ωM is:

$$Z_{MEC} + Z_A = \frac{\omega M}{G(j\omega)(m)} = 72.887 \qquad (7.3)$$

A plot of the resulting combined mechanical and acoustic impedances is shown in Fig. 7.1.

W. H. Watkins, *Loudspeaker Physics and Forced Vibration*,
https://doi.org/10.1007/978-3-030-91634-3_7

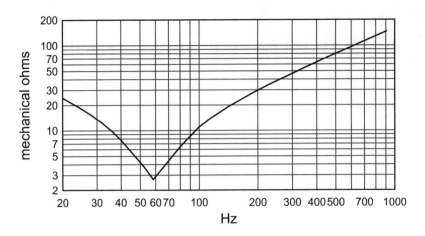

Fig. 7.1 Machanical plus acoustic impedance versus frequency

7.2   Equivalent Mechanical Resistance

The equivalent mechanical resistance is the real part of the mechanical plus acoustic impedance:

$$R_{MEQ} = (Z_{MEC} + Z_A) \cos \phi_{MA} = 9.938 \tag{7.4}$$

in agreement with fundamental physics, where

$$R_{MEQ} = \frac{P_{MEC} + P_A}{\pi \omega A^2 f} = 9.938 \tag{7.5}$$

The equivalent mechanical resistance can also be expressed as the ratio of the component of force doing work to velocity:

$$R_{MEQ} = \frac{F \cos \phi_{MA}}{v} = 9.938 \tag{7.6}$$

The power-to-loss ratio gives R_{MEQ} as

$$R_{MEQ} = \frac{PL}{PL + 1} \frac{P}{v^2} = 9.937 \tag{7.7}$$

From physical principles we also find that

$$R_{MEQ} = \frac{F_{MAX}}{\omega A} \cos \phi_{MA} = 9.9234 \tag{7.8}$$

This last equation is the relationship used in the initial laboratory measurements, and its value was used in the subsequent work presented here.

7.3   Air Load Resistance

The resistance encountered on one side of the cone due to the air load at 456 Hz is [25]:

$$R_{MA} = \frac{\rho_O}{2\pi c} \omega^2 S_D^2 = 2.5312 \tag{7.9}$$

The unit is the MKS mechanical ohm. R_{MA} represents the resistance that power is delivered into to create acoustic power. It is also given with volume acceleration as:

$$R_{MA} = \frac{\rho_O}{2\pi c} \left(\frac{Va}{v}\right)^2 = \frac{P_A}{v^2} = 2.5311 \tag{7.10}$$

7.4   Mechanical Resistance

The mechanical resistance of the combined spider and surround suspensions is

$$R_{MS} = \frac{(Bl)^2}{R_{EMA}} - (2R_{MA}) = 2.5861 \tag{7.11}$$

or with PL (using values at f_O)

$$R_{MS} = \frac{R_{MTR}}{PL} - (2R_{MA}) = 2.5859 \tag{7.12}$$

(Note that R_{MS} is assumed constant with frequency).

Chapter 8
Mechanical and Acoustic Power

8.1 Electrical Power Factor

The electrical power factor is a measure of how effectively the apparent or potential power EI (known as volt-amps) is used. If the load is purely resistive (in this case at f_O and f_E) voltage and current are in phase, the electrical power factor is unity (one), and all the potential power is being used. However if reactance is present in the load at a given frequency, then the voltage and current are no longer in phase and the power factor will be less than one. This indicates that all the potential power is not being used. In the reference driver measurements, the phase angle between the voltage and current at 456 Hz is:

$$\phi_E = arc \; \cos \left(\frac{P}{EI} \right) = 15.11 \text{ deg.} \tag{8.1}$$

or in terms of resistance and impedance,

$$\phi_E = arc \; \cos \left(\frac{R_{ET}}{Z_{ET}} \right) = 15.11 \text{ deg.} \tag{8.2}$$

This phase angle yields an electrical power factor of

$$PF_E = \cos \; \phi_E = 0.9654 \tag{8.3}$$

This represents the extent that voltage is in phase with current. A plot of both electrical and mechanical power factors is shown in Fig. 8.1.

Now the apparent power in volt-amps is:

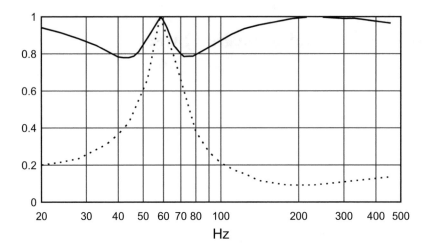

Fig. 8.1 Power factor versus frequency. Electrical (solid). Mechanical + acoustic (dot)

$$S = EI = I^2 Z_{ET} = 0.2947 \qquad (8.4)$$

and the power dissipated is

$$P = EI \cos \phi_E = I^2 R_{ET} = 0.2845 \qquad (5.3)$$

while the reactive power is

$$Q = EI \sin \phi_E = I^2 X_{ET} = 0.0768 \qquad (8.5)$$

The reactive power in effect just shuttles back and forth between the source and load, with no net transfer of power from the source to the load. The apparent power S is denoted by the complex sum of the real and reactive components as $S = P + jQ$.

8.2   Mechanical Power Factor

The mechanical power factor and phase correspond to the electrical side, with force analogous to voltage, and velocity analogous to current. The phase angle between force and velocity is:

$$\phi_{MA} = arc \ \cos \ \frac{P_{MEC} + P_A}{Fv} = 82.16 \text{ deg.} \qquad (8.6)$$

or

$$\phi_{MA} = arc \ cos \ \frac{R_{MEQ}}{Z_{MEC} + Z_A} = 82.17 \ \text{deg.,} \tag{8.7}$$

giving a combined mechanical and acoustic power factor (see Fig. 8.1) of:

$$PF_{MA} = cos \ \phi_{MA} = 0.1363 \tag{8.8}$$

The apparent mechanical plus acoustic power is

$$S_{MA} = Fv = v^2 \left(Z_{MEC} + Z_A \right) = 0.0320 \tag{8.9}$$

and the sum of the mechanical plus acoustic power is

$$P_{MEC} + P_A = (Fv) \ cos \ \phi_{MA} = v^2 R_{MEQ} = 4.366 \ \times \ 10^{-3} \tag{8.10}$$

The reactive power is:

$$Q_{MA} = (Fv) \ sin \ \phi_{MA} = v^2 \left(X_{MEC} + X_{MA} \right) = 0.0317 \tag{8.11}$$

The apparent mechanical plus acoustic power is also $S_{MA} = (P_{MEC} + P_A) + j \ Q_{MA}$. The above holds in the piston-band region up to where the upward slope of the air load resistance levels off, i.e. a frequency of about $f_R = \sqrt{0.5} c / (\pi a_r)$. It is interesting to note that the mechanical plus acoustic power factor is also equal to the ratio of the drive voltage to the motion-generated back emf voltage:

$$PF_{MA} = \frac{E_{DR}}{E_B} = 0.1363 \tag{8.12}$$

Now the relationship to the power-to-loss ratio is:

$$PF_{MA} = \frac{PL}{PL + 1} \frac{E}{E_B} \ cos \ \phi_E = 0.1363 \tag{8.13}$$

This term represents the extent that force is in phase with velocity.

8.3   Mechanical Plus Acoustic Power

With reference to the mechanical and acoustic load, power in equals power out may be written as:

$$I^2 R_{EMA} = \left\{ \left[(v^2 R_{MS}) + (v^2 2 R_{MA}) \right] + j \left[v \left(Ma + \sqrt{0.5}kA \right) \right] \right\} \cos \phi_{MA} = 4.366 \times 10^{-3}$$

$$(8.14)$$

The term on the left is input power driving the mechanical and acoustic load. The first term in brackets is power dissipated into the resistance of the suspensions and air load respectfully, and the second term in brackets is power overcoming the force due to inertia of the load and suspension stiffness. Note the stiffness force of the suspensions kA is negative with reference to the inertial force of the mass Ma. At mechanical resonance these two opposing forces are equal in magnitude, and they cancel out of the equation. What happens at resonance is, during steady-state motion the energy due to the two forces just sloshes back and forth, changing every quarter cycle between stored kinetic energy in the mass and stored potential energy in the spring-like stiffness of the suspensions. Thus at mechanical resonance the two forces alternate in driving each other, and no power from the source is required to accelerate the mass or stretch the suspensions. The only power needed from the source is that required to overcome the frictional resistance of the suspensions and air load. For peak power in Eq. (8.14), the $\sqrt{0.5}$ term may be dropped and peak values obtained.

The electrical input power driving the mechanical and acoustic load may be found as the total power dissipated minus that dissipated in the coil and core:

$$P_{EMA} = P - \left[I^2 \left(R_E + R_{ECOR} \right) \right] = 4.367 \times 10^{-3} \qquad (8.15)$$

or with the power-to-loss ratio and total input power,

$$P_{EMA} = \frac{PL}{PL + 1} P = 4.366 \times 10^{-3} \qquad (8.16)$$

The power on the mechanical and acoustic output side, as previously shown, is:

$$P_{MEC} + P_A = (Fv) \cos \phi_{MA} = 4.366 \times 10^{-3}, \qquad (8.10)$$

otherwise given as

$$P_{MEC} + P_A = v^2 R_{MEQ} = 4.360 \times 10^{-3} \qquad (8.17)$$

This is in agreement with the motor power rule where power equals current times back emf

$$P_{MEC} + P_A = I E_B \cos \phi_{MA} = 4.366 \times 10^{-3} \qquad (8.18)$$

So we thus see that power in equals power out.

A plot of the combined mechanical and acoustic power is shown in Fig. 8.2.

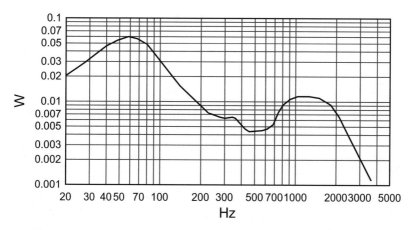

Fig. 8.2 Mechanical plus acoustic power versus frequency

8.4   Relations of Acoustic Power

Now looking at the acoustic side, the acoustic power on one side of the cone at 456 Hz is:

$$P_A = (P_{MEC} + P_A) \, \frac{R_{MA}}{R_{MEQ}} = 1.113 \times 10^{-3} \qquad (5.1)$$

Acoustic power is related to volume acceleration as:

$$P_A = \frac{\rho_O}{2\pi c} \, Va^2 = 1.112 \times 10^{-3} \qquad (8.19)$$

and as related to kinetic energy,

$$P_A = KE_{(MAX)} \, \frac{R_{MA}}{M} = 1.112 \times 10^{-3} \qquad (8.20)$$

and with the power-to-loss ratio,

$$P_A = \frac{PL}{PL+1} \, \frac{R_{MA}}{R_{MEQ}} \, P = 1.113 \times 10^{-3} \qquad (8.21)$$

and with the volume velocity,

$$P_A = \frac{\rho_O}{2\pi c} \, (\omega U)^2 = 1.112 \times 10^{-3} \qquad (8.22)$$

all in agreement with Beranek [14].

$$P_A = v^2 R_{MA} = 1.112 \times 10^{-3} \tag{5.2}$$

A plot of acoustic power as a function of frequency is shown below in Fig. 8.3.

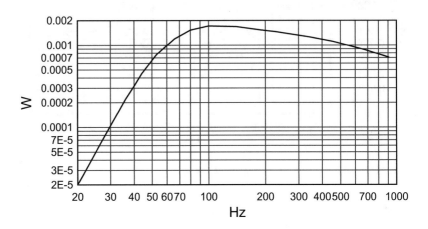

Fig. 8.3 Acoustic power output versus frequency

8.5   Energy and Power Distribution

The energy expended (work done) by the source per cycle on the mechanical and acoustic load at 456 Hz is:

$$W_{EMACY} = I^2 R_{EMA} T = 9.57 \times 10^{-6}; \tag{8.23}$$

the energy expended per cycle heating the coil with a sinusoidal waveform is:

$$W_{ECLCY} = I^2 R_E T = 5.45 \times 10^{-4}; \tag{8.24}$$

the energy expended per cycle heating the core is:

$$W_{ECRCY} = I^2 R_{ECOR} T = 6.97 \times 10^{-5}, \tag{8.25}$$

giving the total energy expended (work done by the source) per cycle of

$$W_{ECY} = W_{EMACY} + W_{ECLCY} + W_{ECRCY} = 6.24 \times 10^{-4} \tag{8.26}$$

or as expressed with total resistance:

$$W_{ECY} = I^2 R_{ET} T = 6.24 \times 10^{-4} \tag{8.27}$$

The power delivered by the source is the total work done per cycle W_{ECY} times the number of cycles per second

$$P = W_{ECY} f = 0.2845 \tag{8.28}$$

This result is in agreement with the power rule $P = EI \cos \phi_E = 0.2845$ watt, as well as that measured with the power analyzer. Power distribution is shown in Table 8.1 and in graphical form in Fig. 8.4. Note that the power dissipated is a maximum at the electrical resonance of 228 Hz. At 20 Hz the core dissipates only 1% of the power, but this increases as frequency increases. At 14,560 Hz the core power

Table 8.1 Power distribution into the core, coil, and mechanical plus acoustic load, along with the percentage into each. There is some error in the highest frequencies due to small numbers and rounding

f	P	P_{COR}	%	P_{COIL}	%	$P_{MEC} + P_A$	%
20	0.2850	0.00287	1.0	0.2618	91.9	0.0203	7.1
29	0.2320	0.00238	1.0	0.1975	85.1	0.0321	13.8
58.3	0.0773	0.00033	0.4	0.0171	22.1	0.0599	77.5
114	0.2540	0.0073	2.9	0.2206	86.8	0.0261	10.2
228	0.3200	0.0198	6.2	0.2929	91.5	0.0073	2.3
456	0.2845	0.0318	11.2	0.2483	87.3	0.0044	1.5
912	0.2125	0.0415	19.5	0.1603	75.4	0.0107	5.0
1820	0.1433	0.0459	32.0	0.0881	61.5	0.0093	6.5
3640	0.0950	0.0473	49.8	0.0466	49.1	0.0011	1.2
7280	0.0640	0.0418	65.3	0.0222	34.7	0.0002	0.3
14,560	0.0435	0.0322	74.0	0.0108	24.8	0.0001	0.2

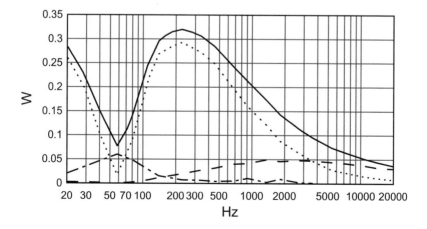

Fig. 8.4 Power versus frequency. P (solid). P_{COIL} (dot). P_{COR} (dash). $P_{MEC} + P_A$ (dot-dash)

dissipation is up to 74.0%. A basic conservation-of-power equation where power in
is equal to power out may be written as:

$$\left(I^2 R_E\right) + \left(I^2 R_{ECOR}\right) + \left(I^2 R_{EMA}\right) = P_{HE} + P_{MEC} + P_A = 0.2845 \qquad (8.29)$$

where the first term on the left is power into the coil, the next is power into the core,
and the last is power driving the combined mechanical and acoustic loads. The first
term on the right represents power converted into heat, the next is mechanical power
produced, and the last is the radiated acoustic power.

Chapter 9
Low-Frequency Response

9.1 Half-Power Point

In the piston-band region above mechanical resonance the speaker's mechanical system is primarily mass-controlled. Looking at say 600 Hz, then with constant power input, as frequency decreases, mass also reactance decreases, and velocity doubles for each lower octave. Now as the frequency is lowered, the air load resistance against the cone also progressively decreases. Thus the increasing velocity offsets the decreasing air load resistance and the resulting acoustic output is relatively flat by nature. Then at mechanical resonance, the mass reactance is in effect canceled out by the stiffness reactance of the enclosed volume of air in the cabinet, and response will be primarily controlled by the effective resistive elements of the system. Then as we move below resonance stiffness reactance is in control, and velocity halves for each lower octave (see Fig. 11.1). This, along with the still decreasing air load resistance, causes acoustic output to decrease by 12 dB with each lower octave. The point below resonance where the acoustic output has dropped to 3 dB below the piston-band level is known as the *half-power frequency*. We now offer some expressions that closely approximate this frequency. A very basic expression is simply a ratio of resistance to mass reactance with allowance for the system Q (using values at f_O):

$$f_3 = \sqrt{Q_T} \, \frac{R_{MTR}}{2 \pi M} = 50.8 \tag{9.1}$$

Note that by omitting $\sqrt{Q_T}$ the above follows Kloss [17] as the ratio of motor resistance to mass. We may also find f_3 as:

$$f_3 = \sqrt{Q_T} \, \frac{f_O}{Q_E} = 50.8 \tag{9.2}$$

or with the power-to-loss ratio and efficiency bandwidth product,

$$f_3 = \sqrt{Q_T} \, PL \, BW \; = \sqrt{Q_T} \, EBP = 50.8 \qquad (9.3)$$

Now using the power-to-loss term and bandwidth gives:

$$f_3 = \sqrt{\left(PL^2/PL + 1\right) f_O \, BW} = 50.8 \qquad (9.4)$$

The measured value by Keele's near-field method [26] was 52.6 Hz, showing the calculated value to be off by 3.4%. The average error of 12 drivers tested was 2.9%, with the calculation too high on 6 units, too low on 4, and correct on 2 (this is with f_3 rounded to the nearest cycle). It should be noted that this measurement gives more variable results than most. This is due to the compliance of the suspensions being quite sensitive to temperature, drive level, and break-in time around the frequency of resonance. Figure 9.1 below depicts the laboratory measurement setup used to confirm the calculated driver parameters.

Frequency response in the piston-band and below, which shows the low frequency roll-off is shown below in Fig. 9.2, measured with the near-field method. The 1.20-dB dip at 283 Hz is attributed to recoil of the motor assembly. For a detailed analysis on motor recoil, see Sect. 11.4 on recoil velocity and Fig. 11.3.

Fig. 9.1 Measuring low-frequency response in the near field

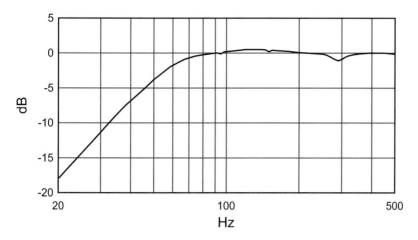

Fig. 9.2 Frequency response (near-field)

9.2   Maximum Amplitude

For Q_T values greater than $\sqrt{0.5}$, the response just above mechanical resonance will rise above that in the piston-band. The increase relative to the piston-band level is:

$$G_{max} = 20 \, \log \frac{1}{2\zeta \sqrt{1-\zeta^2}} = 0.2\,\text{dB} \tag{9.5}$$

The frequency of maximum amplitude is

$$f_{G\,max} = \frac{f_o}{\sqrt{1-(2\zeta^2)}} = 129.0\,\text{Hz} \tag{9.6}$$

A more detailed discussion of the dynamics of low frequency response may be found in Chap. 16.

Chapter 10
Force

10.1 The Lorentz Force Law

Force in linear motors is a push or pull, such that if applied to an object and sufficiently large, causes the object to accelerate. Force is a vector quantity, such that net force, i.e. the component of force F in the direction of displacement that does work, is of magnitude $F \cos \phi_{MA}$. By convention, force is deemed positive if it aids the motion and negative if it opposes the motion. The basic expression for the electromagnetic force driving the mechanical and acoustic load during steady-state motion follows from the Lorentz force law as motor strength times current:

$$F = BlI = 1.5278 \tag{10.1}$$

Newton's well-known second law of motion for force is $F = Ma$. In this case, mechanical resonance serves to decrease force at and around its frequency of occurrence. We may use the mechanical response function to account for this. Fundamental resonance is still having a very slight effect at 456 Hz, such that

$$F = \frac{Ma}{G(j\omega)(m)} = 1.5278. \tag{10.2}$$

The power-to-loss ratio gives

$$F = \frac{F_{MO}}{PL+1} = 1.5276 \tag{10.3}$$

In terms of mechanical impedance and velocity, the force is:

$$F = v\left(Z_{mec} + Z_A\right) = 1.5278 \tag{10.4}$$

© The Author(s), under exclusive license to Springer Nature Switzerland AG 2022
W. H. Watkins, *Loudspeaker Physics and Forced Vibration*,
https://doi.org/10.1007/978-3-030-91634-3_10

10.2   Motor Resistance and Electromagnetic Damping

If the magnitude of generated motor back force due to back emf is given in the form of Bll, we have

$$F_{MB} = Bl \; \frac{E_B}{R_E + R_{ECOR}} = 0.17465 \tag{10.5}$$

Then the motor equivalent mechanical resistance, analogous to $R = E/I$, is:

$$R_{MTR} = \frac{F_{MB}}{v} = 8.3325 \tag{10.6}$$

Substituting Eq. (10.5) for F_{MB}, then Blv for E_B gives

$$R_{MTR} = \frac{Bl \; Blv/(R_E + R_{ECOR})}{v} \tag{10.7}$$

Simplifying this gives:

$$R_{MTR} = \frac{(Bl)^2}{R_E + R_{ECOR}} = 8.3323 \tag{10.8}$$

This represents the equivalent mechanical resistance due to the motor. It is real and produces electromagnetic damping, which is equivalent to mechanical damping. The unit is the MKS mechanical ohm. With the coil blocked for no motion, or at an end-point of travel where the mass has stopped for an instant to reverse its direction of travel, the electromagnetic starting force in the form of Bll and, ignoring reactance, is:

$$F_{ST} = Bl \; \frac{E}{R_E + R_{ECOR}} = 1.6072 \tag{10.9}$$

Then with motion the total applied electromagnetic force becomes $F_{ST} \cos \phi_E$

$$F_{MO} = Bl \; \frac{E}{(R_E + R_{ECOR})} \; \cos \phi_E = 1.5516 \tag{10.10}$$

Olson [27] gives this as:

$$F_{MO} = Bll + \left(\frac{Bll \; R_{MTR}}{Z_{MEC} + Z_A} \; \cos \phi_{MA} \right) = 1.5516 \tag{10.11}$$

Note this is greater than Bll, the force that drives the mechanical and acoustic load. Olson didn't elaborate on the term in the brackets, but we shall find the difference to be the portion of the total applied force overcoming motor resistance. Now the resistance due to the motor $R_{MTR} = (Bl)^2/(R_E + R_{ECOR})$ will cause a retarding electromagnetic damping force F_{DMTR} that opposes the drive force. It may be stated in Ohm's Law form analogous to $E = RI$, otherwise as the effective part of generated motor back force F_{MB}:

$$F_{DMTR} = \frac{(Bl)^2}{(R_E + R_{ECOR})}\, v\, \cos\phi_{MA} = Bl\, \frac{E_B}{(R_E + R_{ECOR})}\, \cos\phi_{MA} = 0.02381$$
$$(10.12)$$

F_{DMTR} is also given by the term in the brackets in Eq. (10.11). Now the source must supply a force of equal magnitude to overcome the opposing motor damping force F_{DMTR}. Let's call the portion of the drive force that overcomes the motor damping F_{ED}. It may be written as:

$$F_{ED} = \frac{PL}{PL+1}\, F_{MO} = F\, \frac{R_{EMA}}{R_E + R_{ECOR}} = 0.02381 \qquad (10.13)$$

If we subtract F_{ED} from the total driving force with motion F_{MO}, then the force left over should be that driving the mechanical and acoustic load

$$F = F_{MO} - F_{ED} = 1.5278 \qquad (10.14)$$

and indeed it is, since we recognize this as $Bll = 1.5278$, the de-facto term for motor force that drives a mechanical load, including of course any air resistance encountered. Since the damping force of the motor opposes the applied driving force, it is actually negative in its function. Then during steady-state motion with the two opposing forces equal in magnitude, they balance out as forces tend to do, such that:

$$F_{DMTR} + F_{ED} = 0 \qquad (10.15)$$

Thus, from a force point of view we see that a portion of the total applied force must be used to overcome the damping due to the motor resistance $(Bl)^2/(R_E + R_{ECOR})$ itself, such that this portion of applied force never gets past the electromagnetic stage to help drive the mechanical and acoustic load. Figure 10.1 shows the relationship of applied electromagnetic force F_{MO}, motor damping force F_{DMTR}, and driving force $F = Bll$.

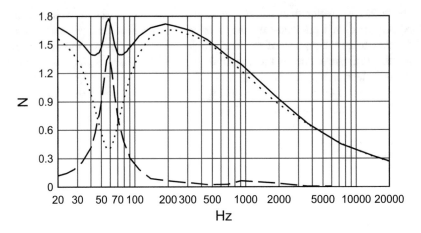

Fig. 10.1 Force versus frequency. F_{MO} (solid), F (dot), F_{DMTR} (dash)

10.3   Correlation with Beranek's Velocity Expression

Beranek's Eq. (7.1) [28] gives a relationship to force, impedance, and velocity as:

$$v \approx \frac{Bl\,E/(R_E + R_G)}{R_M + jX_M},$$ (10.16)

where

$$R_M = \frac{(Bl)^2}{R_G + R_E} + R_{MS} + 2R_{MA}$$ (10.17)

and

$$X_M = \omega M - \frac{1}{\omega C_{MT}}$$ (10.18)

Assuming $R_G = 0$, as we are throughout here, and adding R_{ECOR} to R_E gives:

$$v = \frac{[Bl\,E/(R_E + R_{ECOR})]\,\cos\,\phi_E}{\left\{\left[(Bl)^2/(R_E + R_{ECOR})\right] + R_{MS} + 2R_{MA}\right\} + j(\omega M - 1/\omega C_{MT})} = 0.02090$$ (10.19)

The numerator gives the applied force with motion $[Bl\,E/(R_E + R_{ECOR})]\,\cos\,\phi_E = 1.5516$, which is the same as F_{MO} in both Eqs. (10.10) and (10.11). All three of these drive force expressions include the component F_{ED}, which is the portion of the drive force overcoming the damping resistance of the motor $(Bl)^2/(R_E + R_{ECOR})$. The remainder is the force driving the mechanical and acoustic load. This may be shown by subtracting F_{ED} out of the numerator and $(Bl)^2/(R_E + R_{ECOR})$ out of the denominator:

$$v = \frac{\{[Bl\,E/(R_E + R_{ECOR})]\cos\,\phi_E\} - F_{ED}}{\left\{ Rm - \left[(Bl)^2/(R_E + R_{ECOR})\right]\right\} + j(\omega M - 1/\omega C_{MT})} = 0.02096 \quad (10.20)$$

These velocity results are in excellent agreement with velocity of 0.02098 as measured with the laser vibrometer.

Looking at Eq. (10.19), and noting that at mechanical resonance f_O the reactance of mass and compliance in effect cancel each other out, then the last term in the denominator is zero. This leaves motor resistance $(Bl)^2/(R_E + R_{ECOR})$ to be mainly in control of velocity and thereby acoustic output in this region.

10.4   Relation to the Power-to-Loss Ratio

Relationships with the foregoing and the power-to-loss ratio in magnitude are:

$$F_{ST} = \frac{(PL + 1)\,F}{\cos\,\phi_E} = 1.6072 \quad\quad (10.21)$$

$$F_{MO} = (PL + 1)\,F = 1.5516 \quad\quad (10.22)$$

$$R_{MTR} = PL\,\frac{(Bl)^2}{R_{EMA}} = (PL + 1)\,\frac{(Bl)^2}{R_{ET}} = 8.3323 \quad\quad (10.23)$$

$$F_{MB} = \frac{PL\,F}{\cos\,\phi_{MA}} = 0.1746 \quad\quad (10.24)$$

$$F_{ED} = PL\,F = 0.02381 \quad\quad (10.25)$$

10.5   Force Expressions

During steady-state motion and as related to force equilibrium, an equation of motion may be written such that the applied electromagnetic force F_{MO} is equal in magnitude to the sum of all the damping forces of the load. In *RMS* units this is:

$$F_{MO} = (R_{MTR}\, v\, \cos\, \phi_{MA}) + \left\{ v\left[(R_{MS} + 2R_{MA}) + j\left(\omega M - \frac{1}{\omega C_{MT}}\right)\right]\right\} = 1.5516$$

$$(10.26)$$

where F_{MO} is applied electromagnetic force $[Bl\, E/(R_E + R_{ECOR})]\, \cos\, \phi_E = 1.5516$. The first term on the right side of the equation is the electromagnetic damping force of the motor, and the last term is the retarding force of the mechanical and acoustic load. Now the force equation as related to Newton's second law of motion gives applied force F_{MO} equal in magnitude to the damping forces of the load in *RMS* units as:

$$F_{MO} = (R_{MTR}\, v\, \cos\, \phi_{MA}) + \left\{ [v(R_{MS} + 2R_{MA})] + j\left[(M a) + \left(\sqrt{0.5}kA\right)\right]\right\} = 1.5516$$

$$(10.27)$$

where $k A$ is negative, and $k = 1/C_{MT} = \omega_O{}^2 M$ is the suspension stiffness. The force equation is given in mechanical and acoustic terms with force F driving the mechanical and acoustic load equal in magnitude to the mechanical and acoustic damping forces of the load in *RMS* units as:

$$F = v\left[(R_{MS} + 2R_{MA}) + j\left(\omega M - \frac{1}{\omega C_{MT}}\right)\right] = 1.5278 \qquad (10.28)$$

where F is the applied electromagnetic force $Bll = 1.5278$.

The relation to Newton's second law of motion in magnitudes and *RMS* is:

$$F = [v\, (R_{MS} + 2R_{MA})] + j\left[(M a) + \left(\sqrt{0.5}\, kA\right)\right] = 1.5278 \qquad (10.29)$$

For calculating peak forces, the factor $\sqrt{0.5}$ may be dropped and peak values used. Since the term on the right of the equality sign in Eqs. (10.28) and (10.29) is the retarding force of the load, it is actually negative, such that during steady-state motion the applied and retarding forces balance out, and the effective net force is:

$$F + [v\, (Z_{MEC} + Z_A)] = 0 \qquad (10.30)$$

Looking at this from a velocity point of view, at an end point of travel the coil and cone has stopped motion for an instant to reverse direction of travel. Then motion begins and velocity increases until the magnitude of the negative retarding force has built up to equal that of the driving force. In other words, velocity increases until the applied and retarding forces have reached equilibrium.

Chapter 11
Velocity

11.1   Average and RMS Velocity

Average velocity is defined as the distance traveled divided by the time taken to travel that distance. During a quarter cycle of a sinusoidal drive wave, the moving mass travels from an end point of motion to the center of travel equilibrium position. This distance is referred to as amplitude A. Since the distance traveled and the time taken are the same for each quarter cycle, the average velocity for a given quarter cycle is the same as the overall average velocity. The time taken to travel 1/4 cycle is $T/4$, such that:

$$v_{(AVG)} = \frac{A}{(T/4)} = \frac{4A}{T} \tag{11.1}$$

$T = 1/f$, so that

$$v_{(AVG)} = 4Af \tag{11.2}$$

Conversion to *RMS* gives

$$v = \frac{0.5\,\pi}{\sqrt{2}}\,4Af \tag{11.3}$$

Simplifying gives the *RMS* velocity at 456 Hz of:

$$v = \frac{\omega A}{\sqrt{2}} = \sqrt{0.5}\,\omega A = 0.02096 \tag{11.4}$$

Figure 11.1 is a plot of cone velocity versus frequency, as measured with a Polytec PDV-100 laser vibrometer.

W. H. Watkins, *Loudspeaker Physics and Forced Vibration*,
https://doi.org/10.1007/978-3-030-91634-3_11

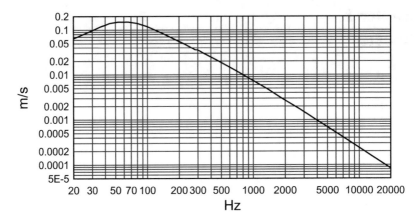

Fig. 11.1 Velocity versus frequency

11.2   Velocity Relationships and Expressions

Logically, velocity is the effective component of force doing work divided by the resistance this force is working against:

$$v = \frac{F}{R_{MEQ}} \cos \phi_{MA} = 0.02099 \tag{11.5}$$

Velocity relates to kinetic energy as

$$v = \sqrt{KE_{(MAX)}/M} = 0.02096, \tag{11.6}$$

and using the power-to-loss ratio yields:

$$v = \frac{PL}{PL+1} \frac{P}{F \cos \phi_{MA}} = 0.02096; \tag{11.7}$$

volume acceleration gives

$$v = \frac{Va}{\omega S_D} = 0.02096, \tag{11.8}$$

and with force and impedance terms we get [30]:

$$v = \frac{BlI}{(Z_{MEC} + Z_A)} = 0.02096 \tag{11.9}$$

Newton's Second Law of Motion may be stated mathematically as:

$$F = M a \tag{11.10}$$

and $a = \omega v$, so that

$$F = M \omega v \tag{11.11}$$

from which, if there were no mechanical resonance,

$$v = \frac{F}{\omega M} \tag{11.12}$$

The parameter Q_M affects velocity profoundly around mechanical resonance, and is still having a slight influence at 456 Hz (about 0.095 dB). Therefore, allowance for the mechanical response function due to Q_M gives:

$$v = \frac{F}{\omega M} \, G(j\omega) \, (m) = 0.02096 \tag{11.13}$$

Velocity may be found from voltage as:

$$v = \frac{(E \cos \phi_E) - [I (R_E + R_{ECOR})]}{Bl \cos \phi_{MA}} = 0.02096. \tag{11.14}$$

11.3   Measured Velocity

As noted, average cone velocity was measured with a Polytec PDV-100 laser vibrometer. Some 68 near-field measurements were taken, one in the center of each of 68 equal-area blocks laid out within the cone area (see Fig. 11.2). The 68 measurements were then averaged. Velocity as used throughout here is true velocity, analogous to the airspeed of an airplane. If an airplane is traveling at 100 mph with reference to the ground and is flying directly into a 5 mph headwind, its true velocity is 105 mph. That relationship is given by:

$$v_{(TRUE)} = v_{(GROUND)} - v_{(WIND)} = 105 \text{ mph} \tag{11.15}$$

where wind velocity $v_{(WIND)}$ is negative with respect to the motion of the airplane.

Now the magnetic field due to drive current through the voice coil works against the magnetic field in the air gap of the motor. At 456 Hz, the average velocity of the cone, and therefore that of the coil as measured with the vibrometer from a stationary ground position was 0.01828 m/s. The average recoil velocity of the pole and plate, and therefore that of the motor's magnetic field, as measured from the same ground position was 6.1×10^{-4} m/s. This motion of the motor's magnetic

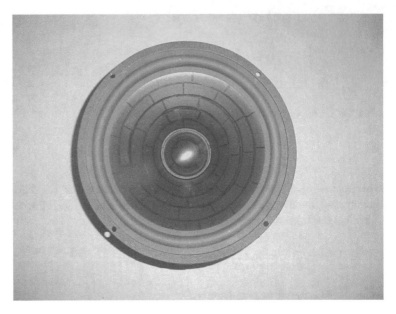

Fig. 11.2 Cone marked into blocks for velocity measurement

field is directly opposite and opposing the coil motion, so the coil is working against the motor's oncoming magnetic field, just as the airplane works against a headwind. This means that true average coil and cone velocity is greater than measured from the ground, such that true average velocity at 456 Hz is:

$$v_{(TRUEAVG)} = v_{(GROUNDAVG)} - v_{(POLEAVG)} = 0.01889 \qquad (11.16)$$

where pole velocity is negative with respect to the motion of the coil. Conversion to *RMS* values gives true *RMS* velocity as measured with the vibrometer of:

$$v = v_{(TRUEAVG)} \left(0.5\pi / \sqrt{2} \right) = 0.02098, \qquad (11.17)$$

in confirmation of the calculated values above.

11.4   Recoil Velocity

If we calculate recoil as the ratio of pole velocity to true velocity, then at 456 Hz,

$$\text{recoil} = \frac{v_{POLE}}{v} = 0.0323 = 3.23\% \tag{11.18}$$

where in terms of *RMS*, $v_{(POLE)} = v_{(POLEAVG)} \left(0.5\pi/\sqrt{2}\right)$. The *RMS* recoil velocity as measured at the pole face is shown in Fig. 11.3.

Note that mechanical resonance of the motor assembly and frame occurs at 283 Hz with *RMS* recoil velocity as measured at the pole face of 9.0×10^{-3} m/s. The true velocity at this frequency is:

$$v = \frac{Bll}{(Z_{MEC} + Z_A)} = 0.03595, \tag{11.9}$$

giving the recoil at 283 Hz as:

$$\text{recoil}_{(283Hz)} = \frac{v_{POLE}}{v} = 0.250 = 25.0\% \tag{11.19}$$

Now 283 Hz is about one note above middle C on the piano, and about the middle of the range of most musical instruments. In view of this, the audibility of the effect from motor recoil would be of interest, but this was not investigated. The heavier the motor and basket, the lower would be the recoil, of course. Although the reference driver has a die-cast metal basket, adding mass to it would certainly reduce motor recoil, but again this was not investigated further.

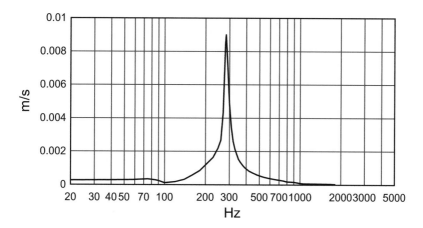

Fig. 11.3 RMS recoil velocity versus frequency

11.5   Volume Velocity

The volume velocity is the rate of flow of the air directly against the area of the cone. It is given for the test unit at 456 Hz as:

$$U = v\, S_D = 4.99 \times 10^{-4} \tag{11.20}$$

or with volume acceleration,

$$U = \frac{Va}{\omega} = 4.99 \times 10^{-4} \tag{11.21}$$

Chapter 12
Energy

12.1 General Considerations

A general definition for energy is the capacity to do work. Potential energy is due to configuration or position, such as a stretched spring, or an object raised above ground. Kinetic energy is the energy an object possesses due to being in motion, which gives it the ability to do work, i.e., transfer its kinetic energy to another object. An example of kinetic energy is a speeding bullet on its way to distort its target. A wrecking ball is an example of both potential and kinetic energy. When raised, it possesses gravitational potential energy due its position above ground. When released, this potential energy is converted into kinetic energy as the ball speeds up. The kinetic energy then does work as the ball hits and breaks loose bricks or whatever. Now as an analogy to the moving mass of the driver, suppose the wrecking ball just swings back and forth without hitting anything. The system energy converts back and forth between kinetic and potential energy, but will gradually dissipate into the resistance of the air and suspension, slowing the ball down. It can be kept up to speed (constant distance of travel per cycle) by adding an amount of energy per cycle equal to that dissipated per cycle. In the case of the driver, the amount of energy dissipated per cycle into the mechanical and acoustic load in the form of work being done is

$$W_{MACY} = \pi \omega R_{MEQ} A^2 = 9.57 \times 10^{-6} \qquad (12.1)$$

Therefore for steady-state motion the source must supply an equal amount of energy per cycle, i.e. do work per cycle on the mechanical and acoustic load of:

$$W_{EMACY} = I^2 R_{EMA} T = 9.57 \times 10^{-6} \qquad (8.23)$$

or

$$W_{EMACY} = PT \, \frac{PL}{PL+1} = 9.57 \times 10^{-6} \tag{12.2}$$

12.2   Kinetic Energy

At an end point of cone travel, velocity and kinetic energy in the system are both zero. Potential energy stored in the system is maximum, since the suspensions are stretched the most they can be. Then with motion, and as the moving mass passes through the center point of travel (the equilibrium position), the suspensions are un-stretched, potential energy is zero, and kinetic energy is maximum, since velocity is at its maximum. The kinetic energy at 456 Hz is then:

$$KE_{(MAX)} = \left(0.5 \, M \, v^2_{FINAL}\right) - \left(0.5 \, M \, v^2_{INITIAL}\right) = 1.13 \times 10^{-5} \tag{12.3}$$

where final velocity $= v\sqrt{2}$. Kinetic energy is also given as

$$KE_{(MAX)} = 0.5 \, M \, \omega^2 A^2 = 1.13 \times 10^{-5} \tag{12.4}$$

Using the power-to-loss ratio gives:

$$KE_{(MAX)} = P \, \frac{M}{R_{MEQ}} \, \frac{PL}{PL+1} = 1.13 \times 10^{-5} \tag{12.5}$$

and solving for mass, resistance, and power gives

$$KE_{MAX} = P_A \, \frac{M}{R_{MA}} = 1.13 \times 10^{-5} \tag{12.6}$$

Figure 12.1 provides a plot of the kinetic energy from 20 through 1000 Hz.

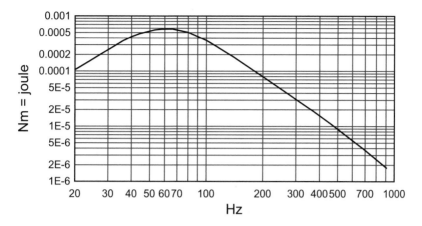

Fig. 12.1 Maximum kinetic energy versus frequency

12.3   Relation of Kinetic Energy to Source Power

The kinetic energy of a moving object is equal to the energy expended on it getting it up to speed. However, energy must also be expended to overcome frictional resistance in the process. The portion of total energy expended by the source on the mechanical and acoustic load that winds up as kinetic energy is energy expended per second (power) on the mechanical and acoustic load times the ratio of mass to equivalent mechanical resistance. At 456 Hz this is:

$$KE_{(MAX)} = \left(I^2 R_{EMA}\right) \frac{M}{R_{MEQ}} = 1.13 \times 10^{-5} \tag{12.7}$$

where the mass term functions as the resistance to acceleration, making the ratio similar to that of a voltage divider, such that it gives the portion of energy transferred that goes into kinetic energy.

Chapter 13
Work, Efficiency, and Power

13.1 Work Done Heating the Coil and Core

Work may be defined as power times the time taken to do the work. Recall that work done per cycle by the source on the mechanical and acoustic load is the electrical power driving the mechanical and acoustic load times the time taken to complete a cycle:

$$W_{EMACY} = I^2 R_{EMA} T = 9.57 \times 10^{-6} \tag{8.23}$$

The work done per cycle by the source creating heat in the coil and core is

$$W_{EHCY} = I^2 (R_E + R_{ECOR}) T = 6.143 \times 10^{-4}, \tag{13.1}$$

giving the total work done per cycle by the source of

$$W_{ECY} = W_{EMACY} + W_{EHCY} = 6.239 \times 10^{-4}, \tag{13.2}$$

which is in agreement with power from the source times time:

$$W_{ECY} = [(EI) \cos \phi_E] T = 6.239 \times 10^{-4} \tag{13.3}$$

13.2 Work Done on the Mechanical and Acoustic Side

Work is also defined as net force times distance traveled. During a full cycle, the distance traveled by the mass is four times the amplitude. This gives work per cycle on the mechanical and acoustic output side of:

© The Author(s), under exclusive license to Springer Nature Switzerland AG 2022
W. H. Watkins, *Loudspeaker Physics and Forced Vibration*,
https://doi.org/10.1007/978-3-030-91634-3_13

$$W_{MACY} = 4AF \cos \phi_{MA} \left(0.5\,\pi/\sqrt{2} \right) = 9.57 \times 10^{-6} \quad (13.4)$$

or as mechanical plus acoustic power times time:

$$W_{MACY} = (P_{MEC} + P_A)\,T = 9.57 \times 10^{-6} \quad (13.5)$$

Thus, the work done per cycle by the source on the mechanical and acoustic load in Eq. (8.23) is equal to the work out. Figure 13.1 shows work done by the source creating heat, work done driving the load, and total work done.

13.3   Relation to Power and the Power-to-Loss Ratio

The overall electrical to acoustic efficiency as related to work is:

$$\eta = \frac{W_{MACY}f}{P}\frac{R_{MA}}{R_{MEQ}} = 0.00391 = 0.391\%. \quad (13.6)$$

Now the concept of "work in equals work out" may be written as:

$$W_{ECY} = \left[(Fv \cos \phi_{MA})\,T \right] + \left[\left(I^2 R_E + R_{ECOR} \right) T \right] = 6.239 \times 10^{-4} \quad (13.7)$$

Power relates to work as work done per cycle divided by the time taken to complete the cycle,

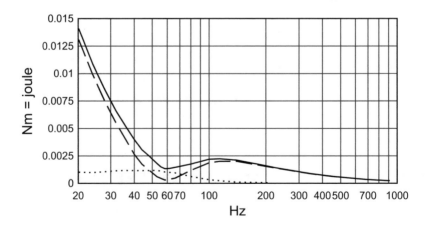

Fig. 13.1 Work versus frequency. Total (solid), creating heat (dash), moving the load (dot)

$$P = \frac{W_{ECY}}{T} = 0.2845 \tag{13.8}$$

Also, since power is defined as work done per second, power is work done per cycle times the number of cycles per second:

$$P = W_{ECY}f = 0.2845 \tag{8.28}$$

This is in agreement with the measured power. The basic conventional expression for power is

$$P = (EI) \cos \phi_E = 0.2845 \tag{5.3}$$

Other, somewhat simpler expressions for work are:

$$W_{ECY} = \frac{P}{f} = 6.239 \times 10^{-4} \tag{13.9}$$

$$W_{MACY} = \frac{(P_{MEC} + P_A)}{f} = 9.57 \times 10^{-6} \tag{13.10}$$

$$W_{EHCY} = \frac{P_H}{f} = 6.14 \times 10^{-4} \tag{13.11}$$

$$W_{EMACY} = \frac{I^2 R_{EMA}}{f} = 9.57 \times 10^{-6} \tag{13.12}$$

Now the power-to-loss ratio relates to the preceding as:

$$W_{EMACY} = \frac{PL}{PL+1} PT = 9.57 \times 10^{-6} \tag{13.13}$$

$$W_{EHCY} = \frac{I^2 R_{EMA} T}{PL} = 6.143 \times 10^{-4} \tag{13.14}$$

$$W_{ECY} = \frac{PL+1}{PL} I^2 R_{EMA} T = 6.24 \times 10^{-4} \tag{13.15}$$

Chapter 14
Resonance, Q, and Measurements

14.1 Mass, Compliance, and Mechanical Resonance

Mass represents the measure of inertia, i.e. the resistance of a body to a change of motion. Since the driver was not measured in a vacuum, the mass of the air load was moved along with that of the cone and coil. Therefore as previously noted, mass as defined here includes the air load mass. Measurement of mass was made per Beranek [31] as:

$$M = \frac{M_{(ADDED)}}{\left(f_O/f_{O(NEW)}\right)^2 - 1} = 0.02572 \tag{14.1}$$

where $M_{(ADDED)}$ is mass added to the cone, and $f_{O(NEW)}$ is the resonant frequency with added mass. Mass may also be found per Newton and with compliance as:

$$M = \frac{F}{a}\, G(j\omega)\,(m) = \frac{1}{\omega_O{}^2\, C_{MT}} = 0.02572 \tag{14.2}$$

or with the power-to-loss ratio PL (values at f_O)

$$M = \frac{R_{MTR}}{2\pi\, BW\, PL} = 0.02572 \tag{14.3}$$

The mass of the mechanical moving assembly excluding the air load on both sides of the cone is:

$$M_{MD} = M - \left[(16/3)\rho_O a_r{}^3\right] = 0.02158 \tag{14.4}$$

Mass is assumed to be constant with frequency.

© The Author(s), under exclusive license to Springer Nature Switzerland AG 2022
W. H. Watkins, *Loudspeaker Physics and Forced Vibration*,
https://doi.org/10.1007/978-3-030-91634-3_14

Compliance is a measure of the system elasticity, or the "give" it has when forced. It is the reciprocal of the system stiffness, and is given as:

$$C_{MT} = \frac{1}{k} = \frac{1}{\omega_0^2 M} = 2.896 \times 10^{-4} \tag{14.5}$$

otherwise with the power-to-loss ratio PL (values at f_O):

$$C_{MT} = \frac{PL}{\omega_0 Q_M R_{MTR}} = 2.896 \times 10^{-4} \tag{14.6}$$

Compliance is assumed to be constant with frequency.

Mechanical resonance occurs at the frequency where the reactances of the mass and compliance are equal in magnitude, i.e. where $2\pi f M = 1/(2\pi f C_{MT})$. At mechanical resonance, the mass reactance and compliance reactance are equal in magnitude but in opposition to each other, therefore they balance out. This leaves the effective net reactance to be zero. Thusly the system is resistance controlled at resonance, and with a closed box system the effective motor resistance $(Bl)^2/(R_E + R_{ECOR})$ is dominant. The frequency of mechanical resonance may be found by convention as:

$$f_O = \frac{1}{2\pi \sqrt{M C_{MT}}} = 58.31, \tag{14.7}$$

in agreement with f_O as measured. Otherwise f_O is given with the power-to-loss ratio and values at f_O as

$$f_O = (PL + 1)\, BW\, Q_T = 58.31 \tag{14.8}$$

or

$$f_O = PL\, BW\, Q_E = 58.31 \tag{14.9}$$

14.2 Bandwidth

Bandwidth may be found from the half-power points below and above resonance f_1 and f_2. These two frequencies are located below and above resonance, where with constant applied voltage the current is $\sqrt{2}$ times the current at resonance, giving bandwidth as:

$$BW = f_2 - f_1 = \frac{(Bl)^2}{2\pi M R_{EMA}} = 16.515 \qquad (14.10)$$

where R_{EMA} at $f_0 = 20.022$.

The relation to the power-to-loss ratio PL (values at f_O) is

$$BW = \frac{f_O}{(PL+1)\,Q_T} = \frac{f_O}{Q_M} = 16.515 \qquad (14.11)$$

14.3   Quality Factors at Resonance

Q_T is generally known as the total or overall quality factor, and it gives the response amplitude at resonance relative to that in the piston-band. Q_M is the mechanical quality factor, i.e. the part of Q_T determined by the mechanical and acoustic damping (note values in this section are those at f_O):

$$Q_M = \frac{\omega_O M}{R_{MEQ}} = \frac{M a}{F} = \frac{f_O}{f_2 - f_1} = 3.531 \qquad (14.12)$$

or with the power-to-loss ratioPL

$$Q_M = Q_T\,(PL+1) = 3.531, \qquad (14.13)$$

Now Q_E is the electrical quality factor, i.e. the part of Q_T determined by the electromagnetic damping

$$Q_E = \frac{\omega_O M}{R_{MTR}} = Q_T\,\frac{E}{E_B} = 1.0222 \qquad (14.14)$$

or with the power-to-loss ratio PL,

$$Q_E = \frac{Q_M}{PL} = 1.0223 \qquad (14.15)$$

Q_T is determined by the total of mechanical, acoustic, and electromagnetic damping:

$$Q_T = \frac{\omega_O M}{R_{MTR} + R_{MEQ}} = 0.7927 \qquad (14.16)$$

or with the power-to-loss ratio PL,

$$Q_T = \frac{Q_M}{PL+1} = 0.7927 \qquad (14.17)$$

Q_T is also given as

$$Q_T = \frac{Q_M \, Q_E}{Q_M + Q_E} = \frac{1}{2\zeta} = 0.7927 \qquad (14.18)$$

Response at resonance in decibels relative to that in the piston-band is given as 20 log Q_T. This gives response of the reference driver at resonance as -2.02 dB. Measurement by Keele's method [26] gave -2.0 dB, in excellent agreement. The frequency response reflecting this was shown earlier in Fig. 9.2.

Chapter 15
Related Parameters and Measurements

15.1 Acceleration

Acceleration is the time rate of change of velocity, i.e. how fast velocity changes. In this case, starting from zero velocity at an end point of travel and with sinusoidal motion, average acceleration is change in velocity (maximum velocity) divided by the time taken to travel 1/4 cycle. This gives *RMS* acceleration at 456 Hz of:

$$a = \frac{\Delta v}{\Delta t} \frac{0.5\pi}{\sqrt{2}} = 60.056 \tag{15.1}$$

where Δv = maximum velocity, i.e. $\sqrt{2}\,v$, and $\Delta t = 0.25\ T$. Substitution of this into Eq. (15.1) and combining constants simplifies to

$$a = 2\pi\,\frac{v}{T} = \omega\,v = 60.056 \tag{15.2}$$

Acceleration according to Newton's Second Law of Motion, and using the mechanical response function to allow for mechanical resonance having a slight effect at 456 Hz is given by:

$$a = \frac{F}{M}\,G(j\omega)\,(m) = 60.056 \tag{15.3}$$

Alternatively, the use of volume acceleration gives

$$a = \frac{Va}{S_D} = 60.056 \tag{15.4}$$

Figure 15.1 shows cone acceleration from 20 through 1000 Hz.

W. H. Watkins, *Loudspeaker Physics and Forced Vibration*,
https://doi.org/10.1007/978-3-030-91634-3_15

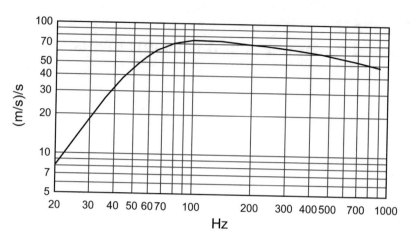

Fig. 15.1 Cone acceleration versus frequency

15.2 Amplitude

Amplitude A is the distance from the equilibrium or rest position, i.e. the center point of travel to an end point of motion. In other words, A represents the distance the mass travels during a quarter cycle. In the most basic form, the distance traveled is average velocity times time. For a quarter cycle, the time taken is T/4, such that

$$A = \frac{F_{(MAX)}}{\omega^2 M} = \frac{v_{(AVG)}}{4} \frac{T}{4} = \frac{v_{(MAX)}}{\omega} = 1.035 \times 10^{-5} \qquad (15.5)$$

Otherwise, using the mechanical response function gives

$$A = \frac{F_{MAX}}{\omega^2 M} MRF = 1.035 \times 10^{-5} \qquad (15.6)$$

The amplitude is shown in Fig. 15.2. Note that with descending frequency it becomes constant from about a half octave below mechanical resonance [32]. This is due to the retarding force (stiffness) of the enclosed air in the cabinet.

Amplitude is also given as

$$A = \frac{F_{(MAX)}}{\omega (Z_{MEC} + Z_A)} = 1.035 \times 10^{-5} \qquad (15.7)$$

or with the power-to-loss ratio

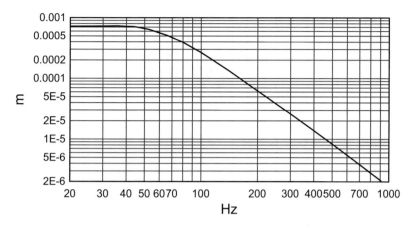

Fig. 15.2 Cone amplitude versus frequency

$$A = \frac{PL}{PL+1} \frac{PT}{4F \cos \phi_{MA}} \frac{\sqrt{2}}{0.5\pi} = 1.035 \times 10^{-5} \qquad (15.8)$$

15.3   SPL

SPL as used here is the free field sound pressure level at 1 m on axis, and radiating into 180°. At 456 Hz, and using volume acceleration, this is found as:

$$SPL = 20 \log \left(Va \frac{\rho_O}{2\pi * 2.0^{-5}} \right) = 82.56; \qquad (15.9)$$

otherwise by convention as

$$SPL = (10 \log P_A) + 112.1 = 82.56 \qquad (15.10)$$

SPL is then given with the power-to-loss ratio as

$$SPL = 10 \log \left(\frac{PL}{Pl+1} P \frac{R_{MA}}{R_{MEQ}} \right) + 112.1 = 82.57 \qquad (15.11)$$

Note we are using 1.41 *RMS* volts input throughout here. The general reference input for *SPL* is 2.83 volts, which would increase *SPL* by 6 dB, to 88.56 dB. Assuming 2.83 volts input, the average *SPL* of 12 quality drivers tested would thusly have been 88.66 dB. This shows the reference driver to be typical of quality eight-inch woofers in sensitivity.

15.4   Motor Strength

Motor strength or force constant is flux density times the length of wire in the magnetic field. It is given in terms of back emf as:

$$Bl = E_B \, \frac{\omega}{a} = 7.310 \tag{15.12}$$

or otherwise with the power-to-loss ratio PL (values at f_O),

$$Bl = \frac{Ma}{(PL+1)\,I\,Q_T} = 7.310 \tag{15.13}$$

15.5   Measuring Bl with Precision

The assembly for measuring Bl is shown in Figs. 15.3 and 15.4. It should be of very stable construction, and the two pulleys must rotate with *very* low friction. The speaker end of the string is attached to a copper *l*-shaped hook that attaches to the dome cap or cone at the edge of the cap. The other end is attached to a ferrite bar (inductor core) about ¼" by 1¼" and dangles about $^1/_3$ of the way into the coil. The inductance of the coil is read out on an accurate inductance meter. For the test driver,

Fig. 15.3 Assembly for measuring Bl

Fig. 15.4 Overall set-up for *Bl* measurement

a General Radio 1657 Digibridge was used. Upward pull on the cone is about 18 g, in effect offsetting gravity and pulling the average 8"speaker cone back to very near its equilibrium center position. Weights placed on high compliance speakers can be around 0.075 kg for 8" units, 0.100 kg for 10" and 0.125 kg for 12". Brass nuts or marbles work well. Whatever used must be non-magnetic, and should be placed evenly around the dome.

A 6-volt lantern battery is initially connected in series with the potentiometers, amp meter, and voice coil. The polarity of the battery voltage is chosen such as to bring the cone up with lowering of resistance. To start the test, we first leave the circuit open (one clip lead unattached to one terminal of the battery works well.) Next, we set the potentiometers to maximum resistance to avoid a jerk when connecting the clip to the battery later. Push the cone down ¼ inch or so and GENTLY let it come back to neutral. The meter should show a stable reading after a few pushes. Now note the inductance reading. This is the target reading to achieve after adding the weight. Add the weight gently, connect the battery, and carefully adjust the pots. To avoid overshoot, use the higher resistance pot first, then when near the target reading, slowly adjust the lower resistance pot. If you overshoot and then come back down, the reading will be off slightly by the amount of difference in up and down settling position (hysteresis). Therefore if you overshoot the target, repeat the process. Make note of the current, repeat several times, and average. With careful technique, the repeatability of each individual measurement will be almost perfect. Calculate *Bl* so

$$Bl = 9.8 \; \frac{mass_{(ADDED)}}{I} = 7.311 \tag{15.14}$$

A *Bl* of 7.311 was the result with the test driver, where the mass added was 0.0746 kg, and the current required to return the cone to the original position was 0.10 amp.

15.6   Momentum

Momentum may be defined as the ability of a moving body to maintain its state of motion. It is the product of mass and velocity:

$$p = Mv = 5.39 \times 10^{-4} \tag{15.15}$$

15.7   Pressure

The acoustic pressure at 1 m, on axis, and radiating into 180°, free field is:

$$pr = \frac{\rho_O}{2\pi} \, Va = \rho_O f \, U = 0.2684 \tag{15.16}$$

or with the power-to-loss ratio *PL*,

$$pr = \frac{PL}{PL + 1} \frac{P S_D f \rho_O}{F \cos \phi_{MA}} = 0.2684 \tag{15.17}$$

Chapter 16
Circumventing Efficiency and Bass Extension Limitations

This chapter contains excerpts from a previous work [29].

16.1 Efficiency and Bass Extension Limitations

A limitation with conventional loudspeaker design is that the laws of physics determine the maximum piston-band efficiency and low-frequency extension that can be obtained in a closed-box loudspeaker system with a given cabinet volume [17]. If more efficiency is desired, bass response must be sacrificed, and conversely to obtain more bass, it is necessary to sacrifice efficiency. The purpose of this article is to describe a method that circumvents this limitation. It is both simple to implement and requires no additional power from the amplifier.

16.2 Back EMF

Since the above restraint is related to the motion-induced back emf generated by the speaker at fundamental resonance, let's look at its makeup. The magnet-voice coil assembly of all dynamic speakers of the type we are discussing is a reciprocating motor. Faraday's law states that an induced voltage will be generated in an electrical conductor when it moves through a magnetic field. This means that during operation the motor generates a voltage of its own. This generated voltage is generally termed back emf, and its polarity is always opposite to that of the externally applied (sinusoidal) voltage; thus, it opposes the applied voltage. This raises the speaker's impedance and acts to reduce current flow through the voice coil.

Now since the back emf is due to the motion, i.e., velocity of the voice coil in the magnetic field, let's look at how velocity and back emf behave. In a sealed enclosure above the region of fundamental resonance, both cone velocity and excursion are

W. H. Watkins, *Loudspeaker Physics and Forced Vibration*,
https://doi.org/10.1007/978-3-030-91634-3_16

low, the air in the box is compressed and rarefied very little, and the effect of its reactance is negligible. Since back emf is proportional to velocity and velocity is low, this means that back emf is also low; therefore, it is not a major factor in determining current flow within the piston band of operation. If we start at 500 Hz, then as we come down in frequency the mass of the cone and coil assembly is easier to move due to decreasing mass reactance, and velocity inherently doubles with each lower octave. Noting that the back emf $E_B = Blv \cos \varphi_{MA}$, then the effective back emf also increases with decreasing frequency. Then as we move on down in frequency and approach fundamental resonance, the air in the box enters the picture, and the moving mass begins to resonate with the compliance of the air in the box. When resonance is reached, the moving mass is at any instant supplying energy to the air in the box (compressing or rarefying it) due to its kinetic energy, or the air in the box is supplying energy to the moving mass (forcing it in or out) due to its potential energy from pressure or rarefaction. Starting from the center point of motion (the cone and coil rest position) the air in the box is neither compressed nor rarefied. Then as the mass moves inward it will compress the air in the box. At the end point of motion the mass stops and the potential energy from the compressed air in the box forces the cone to travel back toward the center. When it reaches the center point of motion the air in the box is again neither compressed nor rarefied, but the kinetic energy in the moving mass (now traveling at maximum velocity) causes the mass to overshoot the center point, now rarifying i.e., "stretching" the air in the box. When the mass reaches the other (outward-most) end point of travel, it stops and the air in the box is now rarefied to its maximum, and its potential energy starts pulling the mass back toward the center. When the mass reaches the center point of motion it is again at maximum velocity, and has maximum kinetic energy. This causes it to continue onward, and the process repeats itself. In fact, if there were no resistance in the system, this process would continue on forever with no external power being applied, and we would have a perpetual motion machine. So with kinetic energy and potential energy being equal and driving the mass back and forth, we could say the mass gets a free ride at resonance. The end effect here is that the reactance from the moving mass is now in effect canceled out or neutralized and velocity is at maximum. With velocity at maximum, back emf is also at maximum and this raises the speaker's impedance significantly. If the voice coil DC resistance is 4 Ω in a typical speaker, back emf will add about 1 Ω at the resistive point above resonance, for a total of about 5 Ω impedance. Now at resonance back emf may add 15 Ω or more, and the total impedance will be several times that above resonance. This effect is shown in the plot of Fig. 16.1.

This rise in impedance greatly reduces an amplifier's ability to deliver power in the low bass region. In fact, a typical speaker system will accept only about ¼ as much power at mechanical resonance as it does in the piston bands mid range. Power input to the speaker at mechanical resonance (58.3 Hz) is:

$$P_{fo} = \frac{E^2}{Z_{ET}} \cos \varphi_E = 0.077 \text{ watts} \qquad (16.1)$$

Whereas at 456 Hz in the piston band, power input is

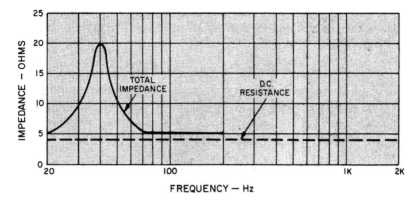

Fig. 16.1 Typical impedance curve. The total impedance is the d.c. resistance plus the reactance due to the back EMF

$$P_{456} = \frac{E^2}{Z_{ET}} \cos \varphi_E = 0.284 \text{ watts} \tag{16.2}$$

So we see that while back emf was not a major factor in determining current flow and wattage intake above resonance, it has raised the impedance and is the dominant factor in determining current flow and wattage intake from an amplifier at and around resonance. In fact, if the power intake at resonance could be brought up to match that above resonance, the bass output would increase by about four times, i.e. be elevated about 6 dB above the higher-frequency output. Then if the output at these higher frequencies above resonance could also be increased, the net result would be both higher efficiency and concurrently overall higher acoustic output. Note the technique to be described here to accomplish this would not require extra power capability from an amplifier. For instance, if an amplifier is rated at 100 watts from 20 through 20,000 Hz, it would be able to deliver its full 100 watts in the bass range instead of being limited to about 25 watts. The process to be described here simply utilizes the existing power that is available in a given amplifier, but not being used before in the low frequency bass range.

16.3   Dominant Parameters

While the device to be described may be characterized in terms of Q_T (the shape of response curve at resonance), the approach used here offers a better depiction of the principle involved. The reasonable assumptions made are that the inductance of the voice coil below 200 Hz is negligible (as in conventional designs), that the amplifier source resistance is negligible (i.e., a damping factor of 25 or more), and that the mechanical and acoustic resistive elements are properly controlled, again as in conventional design. Since acoustic power is determined by velocity-squared times radiation resistance, and back emf is determined by velocity times motor strength Bl, we will work mainly with the velocity term. At 456 Hz the velocity is:

$$v_{456} = \frac{F}{(Z_{MEC} + Z_A)} = 0.0210 \tag{16.3}$$

where F = force applied to the mass assembly and $Z_{MEC} + Z_A$ = mechanical plus acoustic impedance.

Now force is defined as BlI, which at 456 Hz is:

$$F_{456} = BlI = 1.528 \tag{16.4}$$

where

B = flux density in the magnetic gap
l = length of wire in the magnetic gap
I = current through the voice coil

Note that Bl is the motor strength.
Therefore Eq. (16.3) may be written as:

$$v_{456} = \frac{BlI}{(Z_{MEC} + Z_A)} = 0.0210 \tag{16.5}$$

The total electrical impedance may then be written (at 456 Hz) as:

$$Z_{ET} = (R_E + R_{ECOR}) + \frac{(Bl)^2}{(Z_{MEC} + Z_A)} \cos \varphi_{MA} + jX_{ET} = 6.746 \tag{16.6}$$

where $R_E + R_{ECOR}$ is the resistance of the voice coil plus that of the core due to coupling effect.

Now above resonance $R_E + R_{ECOR}$ in Eq. (16.6) dominates, and the current at 456 Hz is:

$$I_{456} = \frac{E}{(R_E + R_{ECOR})} = 0.219 \tag{16.7}$$

Substituting for I in Eq. (16.5), the velocity at 456 Hz is:

$$v_{456} = \frac{Bl[E/(R_E + R_{ECOR})]}{(Z_{MEC} + Z_A)} = \frac{Bl\,E}{(R_E + R_{ECOR})(Z_{MEC} + Z_A)} = 0.022 \tag{16.8}$$

Now at resonance in Eq. (16.6) $[(Bl)^2/(Z_{MEC} + Z_A)] \cos \varphi_{MA}$, the motional impedance term which is due mainly to back emf dominates; therefore the current is:

$$I_{fo} = \frac{E}{(Bl)^2/(Z_{MEC}+Z_A)} = \frac{E(Z_{MEC}+Z_A)}{(Bl)^2} = 0.070 \qquad (16.9)$$

Note that at mechanical resonance $\cos\varphi_{MA} = 1$, this term drops out of the equation.

Substituting for I in Eq. (16.5), the velocity at resonance is:

$$v_{fo} = \frac{Bl\left[E(Z_{MEC}+Z_A)/(Bl)^2\right]}{(Z_{MEC}+Z_A)} = 0.192 \qquad (16.10)$$

Simplifying this gives velocity at resonance simply as:

$$v_{fo} = \frac{E}{Bl} = 0.192 \qquad (16.11)$$

Now if we find the ratio of equivalent electrical resistance R_{EMA} which is due to the driver's mechanical and acoustic impedance that opposes velocity, to the total system resistance R_{ET}, we can then multiply Eq. (16.11) by this ratio to find the exact velocity at mechanical resonance:

$$v_{fo} = \frac{E}{Bl}\frac{R_{EMA}}{R_{ET}} = 0.150 \qquad (16.12)$$

Using the previously developed power-to-loss ratio term gives the velocity at mechanical resonance as:

$$v_{fo} = \frac{E}{Bl}\frac{PL}{PL+1} = 0.150 \qquad (16.13)$$

At frequencies other than mechanical resonance we must deal with reactance. Therefore we may use impedances in the form of $Z = R + jX$ that include both resistance and reactance to obtain the exact magnitude of velocity at any frequency. For example, at 456 Hz velocity is:

$$v = \frac{E}{Bl}\frac{Z_{EMA}}{Z_{ET}} = 0.021 \qquad (16.14)$$

Otherwise we may account for reactance with the phase angle, and subtract out the non-velocity related IR voltage drop $I(R_E + R_{ECOR})$ across the voice coil and core. At 456 Hz this gives:

$$v = \frac{(E \cos \varphi_E) - [I(R_E + R_{ECOR})]}{Bl \cos \varphi_{MA}} = 0.021 \qquad (16.15)$$

The power-to-loss ratio term relates E/Bl to velocity at any frequency as (which at 456 Hz is):

$$v = \frac{E}{Bl} \, PL \, \frac{R_E + R_{ECOR}}{Z_{ET} \cos \varphi_{MA}} = 0.021 \qquad (16.16)$$

or with just the power-to-loss ratio and phase:

$$v = \frac{E}{Bl} \frac{PL}{PL+1} \frac{\cos \varphi_E}{\cos \varphi_{MA}} = 0.021 \qquad (16.17)$$

The mechanical plus acoustic impedance $Z_{MEC} + Z_A$ varies with frequency but is a minimum at mechanical resonance f_o. Equation (16.6) shows that this low mechanical impedance at resonance is reflected back into the electrical circuit by the back emf as a higher impedance. Therefore as noted before, the voice coil does not absorb nearly as much power at resonance as it does above and below resonance. With input voltage E and coil resistance R_E constant, and with a given set of parameters determining $Z_{MEC} + Z_A$ at and above resonance, the following relationships will prevail: A larger motor Bl will increase velocity and output above resonance per Eq. (16.8), and decrease velocity and output at resonance per Eq. (16.9). Conversely, a smaller motor Bl will decrease velocity and output above resonance, and increase velocity and output at resonance. We refer to this behavior as the see-saw effect. Figure 16.2 shows this effect in graphic form where it can be seen that for a given speaker system, i.e. a given cone area, moving mass, and cabinet volume there will be an optimum value, neither too small nor too large, of Bl product (motor strength) that will give the flattest overall response. The results shown in Fig. 16.2 are quite similar to what would be obtained by de-normalizing a set of standard Q_T curves of about 0.75 and 1.5 as shown in their usual normalized format in Fig. 16.3. Q_T curves are often referred to in depicting bass response, however, it should be noted they have usually been normalized with respect to the actual flat band efficiency of each. That is, the zero dB point on the vertical axis is redefined for each speaker regardless of actual efficiency. This normalization process however, has a disadvantage here. It doesn't clearly show that with other things being equal, then when adjusting the motor's Bl product (motor strength), a speaker with a larger motor will be more efficient and have less bass response than a speaker with a smaller motor.

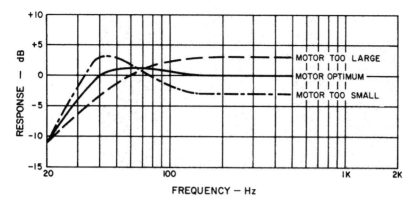

Fig. 16.2 Response for different values of motor strength in the region of fundamental resonance with conventional design

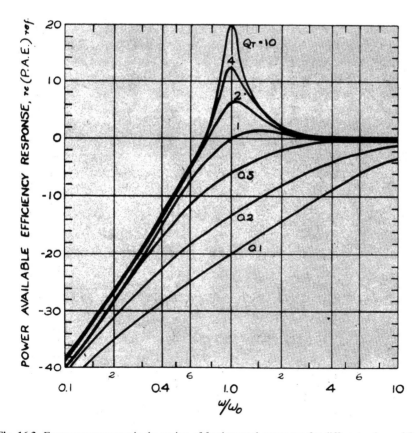

Fig. 16.3 Frequency response in the region of fundamental resonance for different values of Q_T

16.4   The Conventional Trade-Off

From the above considerations, we see that for flat response with a given cone area, moving mass, and cabinet volume we need a motor that's neither too small nor too large. We also see that the often-made assumption that the speaker with the largest magnet is best is not based on the actual facts. When adjusting motor strength then, there is an inevitable trade-off between bass response and efficiency in the flat band. Thus in the final analysis, the crux of the problem is due to the see-saw effect, of varying motor strength. For good efficiency above resonance we need a large motor, while for good bass efficiency we need a smaller motor.

16.5   Dual Motors

The solution is to provide the speaker with two separate motors. This would cut loose the see-saw effect, and we would be free to adjust upper-range response and near-resonance bass response separately, and without one significantly affecting the other. The usual practice in adjusting motor strength is to vary the magnetic strength, i.e. the B in the Bl product. To construct a speaker with two different magnetic fields to drive the same cone would be both impractical and expensive to build, along with other drawbacks.

Let's look at the Bl product in a different way. In a conventional speaker, it is constant with frequency. Suppose we could in effect make l vary with frequency, such that changing the value of Bl in one frequency range would not affect Bl in another range. Figure 16.4 shows a simple method of doing this. A second voice coil is wound over the conventional voice coil and is driven by a series LC resonant circuit, which is adjusted to resonate at the same frequency as the mechanical resonant frequency of the woofer. The electrical circuit depicting this is shown in Fig. 16.5. With proper values of the inductor L and capacitor C, the LC circuit presents almost zero impedance at resonance and a sufficiently high impedance on either side of resonance to effectively remove the second voice coil from the circuit. The second coil has less wire length l than the conventional coil, thus it generates less back emf and has a much lower impedance. This allows more current flow at resonance and increases acoustic output. The electrical Q, and therefore the frequency spread of the dual-drive effect may be controlled in a straightforward manner by adjusting the LC ratio. Taking our example from before, wire length l may be adjusted to lower impedance at resonance to say, 5 Ω (compared to 20 Ω before). This increases the wattage intake to equal that in the piston band, and raises the acoustic output in the bass by 6 dB. Again, note that no extra amplifier power is required. We are simply using the wattage in a given amplifier that was there all the time, but not being utilized in the low bass range by conventional single-coil woofers.

Fig. 16.4 Dual-drive voice coil, showing the return winding on the low impedance coil

Fig. 16.5 Circuit for a dual-motor woofer. L and C are adjusted to resonate at the same frequency as the fundamental acoustic resonance of the woofer. The tuned bandpass circuit formed by L and C allow the secondary voice coil to operate only in the region of resonance

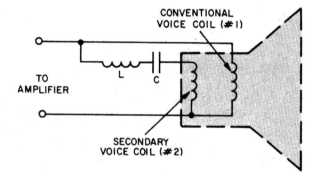

Once it is realized that motor strength can be increased to the point of gap saturation with the attendant over-damping of bass compensated for by the second voice coil, the dividends of the design are apparent. As the flux density is increased, the impedance at resonance is increased, and even fewer turns may be used in the second voice coil without the impedance falling to an unacceptable value. More overall efficiency, a smaller enclosure, extended bass, or a combination of these may be achieved, depending on how the parameters of motor strength, cone area, weight of the moving mass, and enclosure volume are chosen.

The additional cost of building a woofer of this type over a conventional woofer is a pleasant surprise. A magnetic gap of twice the width is not required to accommodate the second voice coil (a wider gap requires a larger magnet to obtain a given flux density). The outer and inner clearance spaces are the same as usual, such that the magnetic gap width needs to be increased only around 25–30%. Most all the expense of a larger magnet will provide benefits that can be put to good use.

Several variations of the principle involved were implemented. While using the circuit of Fig. 16.5, if the main voice coil is in effect disconnected at resonance, a further increase in acoustic output occurs. The back emf from the main coil is much higher than from the low impedance coil, and inserting a parallel LC circuit in series with it in effect takes it out of the circuit at resonance and provides even more acoustic output. Mutual coupling of the two inductors in the two LC circuits can also be accomplished, along with other configurations.

16.6   Technical Details

Figure 16.6 shows the frequency response of a woofer in which a gain of 5½ dB at 30 Hz was achieved. In all the tests the speaker was sunk in the ground with the front edge of the cabinet flush with the ground surface, and well away from any buildings or obstructions (180° solid angle, free-field conditions). The microphone was at 1 m distance, and on-axis with the geometric center of the speaker. The woofer used for the data here was a 12-inch unit. It was installed in a two-cubic-foot cabinet filled with uncompressed R-19 fiberglass. Primary emphasis was placed on extending the bass, very slight emphasis on efficiency, and none on reducing enclosure volume. Taking 87 dB SPL (referenced to 0.0002 microBar), which was produced with a 1-watt input (2.0 volts into a rated impedance of 4 Ω) as reference, the response is flat to 30 Hz, and down only 3 dB at 25 Hz. The response is flat ±1½ dB from 27 through 600 Hz. The frequency response shown in Fig. 16.6 is unique, considering the 2-cubic-foot cabinet volume, sensitivity of 87 dB, response flat down to 30 Hz, and that no equalization or extra amplifier power was used. The design also presents

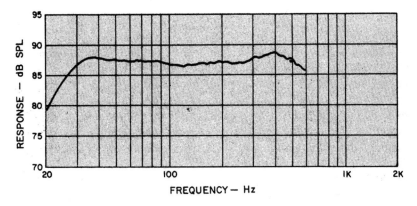

Fig. 16.6 Frequency response of a dual-motor (1 watt into rated impedance of 4 Ω), mike at 1 m on-axis, 180° radiation angle, and free-field conditions. SPL ref. is 0.0002 μ Bar

a far less reactive load, with almost no phase shift, to the amplifier. This almost purely resistive load insures excellent compatibility with virtually any amplifier.

The total harmonic distortion of the test unit is shown in Fig. 16.7. Note that the curves for distortion are for constant sound pressure levels, not constant power input. A specification showing low distortion down to 30 Hz with constant power input is not meaningful, if the acoustic output of the speaker is down 10 dB at this frequency.

Figure 16.8 is a photo of the acoustic output at 30 Hz at 97 dB SPL and with 4% total harmonic distortion. The waveform shows little if any visible distortion at this relatively high level of acoustic output. The impedance for a complete system incorporating the dual-drive woofer is shown in Fig. 16.9. In one prototype, the response curve was almost ruler flat below 200 Hz; however, the best overall results were obtained with slight deviation from this ideal, and the curve shown is the one obtained from the design giving the other data presented here. The minimum impedance is 4.2 Ω at 25 Hz, the maximum is 7 Ω at about 2200 Hz, and an average 4-Ω rating is appropriate. It does not fall under 5 Ω below 20 Hz, and above 20,000 Hz there is a gradual rise in the high end due to the usual inductance of a

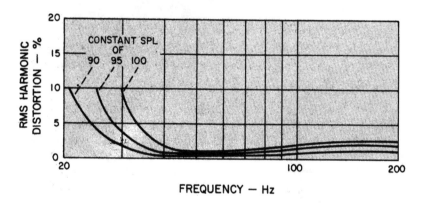

Fig. 16.7 Total rms harmonic distortion with free-field conditions as noted previously. The sound pressure level was held constant for each curve, and the frequency lowered until the distortion reached 10%

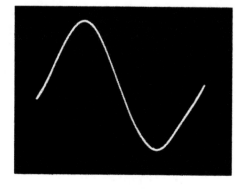

Fig. 16.8 Acoustic output of the woofer at 30 Hz and 97 dB SPL. Free-field conditions as noted previously. (Oscilloscope of microphone output)

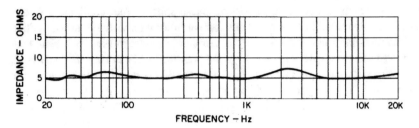

Fig. 16.9   Impedance of a system using dual-motor woofer

Fig. 16.10   Acoustic output
of the woofer in response to
the step-front of a
low-frequency square wave

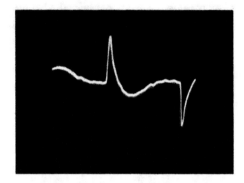

tweeter's voice coil. Note that the conventional peak in impedance at the woofer's mechanical resonance is missing, due to the effect of the second voice coil.

Transient response refers to the ability of the woofer to follow the starting and stopping (attack and decay) of sudden changes in the electrical signal. In a woofer this is a function of the system Q. It is generally considered that a Q of one offers the flattest overall response with a minimum of ringing, and the response curve of Fig. 16.6 conforms to this shape. Since the woofer is driven by a series resonant LC circuit, we need to see if the transient response still follows the rules. A system in a sealed box with a Q of one should have about 30% overshoot to a step-signal, and should decay within one cycle. Figure 16.10 is an oscilloscope photo of the waveform of the acoustic output in response to the steep front of a low-frequency square wave. The overshoot past the neutral or start position is exactly that expected with a Q of one, and shows that the woofer is properly damped both acoustically and electrically. This test was made indoors with the microphone in the near field, and the small ripples in the waveform are due to room boundary reflections.

The frequency response of a complete system consisting of the 12-inch dual-drive woofer, a 4½-inch mid-range driver, and a 1-inch dome tweeter is shown in Fig. 16.11. One-third-octave bands of pink noise were used to produce the curves. The dashed line is the on-axis response, and the solid line is the average of five frontal curves, one on-axis and four at 45° from on-axis, in both vertical and horizontal directions. The response above 3000 Hz could probably benefit from a

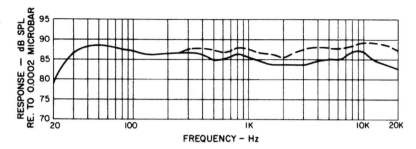

Fig. 16.11 Frequency response of a three-way system using the dual-motor woofer with one-third-octave random noise. Speaker radiating into 180° free-field conditions as noted, with one-watt input

couple of ohms resistance to the tweeter, depending on the characteristics of the listening room.

16.7 Summary of This Chapter

The dual-drive design as just described offers extended bass, higher efficiency, a smaller enclosure, or a combination of these benefits. It does not involve a trade-off in any area of performance, requires no additional amplifier power or equalizer, presents an almost purely resistive load to the amplifier, has the advantage of simplicity of construction, is inexpensive to implement, and offers improvements in performance that are very readily discernible both by measurement and listening. The novel technique was awarded U.S. Patent #3838216, "Device to Effectively Eliminate the Motion-Induced Back EMF in a Loudspeaker System in the Region of Fundamental Acoustic Resonance", granted September 24, 1974. The concept was utilized by Watkins Engineering in several system designs and was also subsequently licensed to Infinity Systems in California for use in several of its high-end speaker systems as the widely praised "Watkins Dual-Motor Woofer".

Appendix A: Parameter Magnitudes and Parameter Tables

Parameter Magnitudes

Magnitudes are those measured at 456 Hz unless noted otherwise. Units are in the standard MKS system.

$$A = (F_{(MAX)}/\omega^2 M)\, MRF = v_{(MAX)}/\omega = 1.0346 \times 10^{-5}(m)$$
$$a = (F/M)\, MRF = Va/S_D = 60.056(m/s^2)$$
$$a_r = \sqrt{S_D/\pi} = 0.087(m)$$
$$Bl = F/I = \omega E_B/a = 7.310(Tm)$$
$$(Bl)^2 = 53.436(Tm * Tm)$$
$$BW = f_2 - f_1 = f_0/[(Pl + 1)\, Q_T] = 16.515(Hz)$$
$$C_{MT} = 1/(\omega_0^2 M) = 2.896 \times 10^{-4}(m/N)$$
$$E = I Z_{ET} = 1.410(V)$$
$$E_B = Blv = 0.1532(V)$$
$$E_{BEFF} = Blv \cos \phi_{MA} = 0.0209(V)$$
$$EBP = PL\, BW = f_0/Q_E = 57.045$$
$$E_{DR} = I R_{EMA} = 0.0209(V)$$
$$E_H = I\,(R_E + R_{ECOR}) = 1.3403(V)$$
$$f_3 = \sqrt{Q_T}\, BW\, PL = 50.79(Hz)$$
$$f_E = f_{ZMIN} = 228(Hz)$$
$$f_O = 1/(2\pi \sqrt{M\, C_{MT}}) = 58.31(Hz)$$
$$F_{ST} = Bl\, E/(R_E + R_{ECOR}) = 1.6072(N)$$
$$F_{MO} = (PL + 1)\, F = F_{ST} \cos \phi_E = 1.5516(N)$$
$$F = BlI = M\, a/MRF = F_{MO} - F_{ED} = 1.5278(N)$$
$$F_{ED} = PL\, F = F R_{EMA}/(R_E + R_{ECOR}) = 0.0238(N)$$
$$F_{DMTR} = R_{MTR}\, v \cos \phi_{MA} = 0.0238(N)$$
$$F_{MB} = Bl\, E_B/(R_E + R_{ECOR}) = 0.1746(N)$$
$$f_{ZMIN} = f_E = 228(Hz)$$

$I \;=\; E/Z_{ET} \;=\; 0.209(A)$

$I_B \;=\; E/Z_{ET} \,(coil\,blocked) \;=\; 0.2067\,(A)$

$k \;=\; 1/C_{MT} \;=\; \omega_O{}^2 M \;=\; 3453(N/m)$

$KE_{(MAX)} \;=\; (P_{MEC} + P_A)\, M/R_{MEQ} \;=\; 1.13 \times 10^{-5}(J)$

$L_E \;=\; [E/(I\omega)]\,\sin\phi_E \;=\; 6.14 \times 10^{-4}(H)$

$L_G \;=\; assumed\;0(H)$

$M \;=\; (F/a)\,MRF \;=\; 1/(\omega_O{}^2 C_{MT}) \;=\; 0.02572(kg)$

$M_{MD} \;=\; M - (16\rho_O a_r{}^3/3) \;=\; 0.02158(kg)$

$MRF \;=\; M\,a/F \;=\; G(j\omega)\,(m) \;=\; 1.0110$

$P \;=\; E\,I \cos\phi_E \;=\; 0.2845(W)$

$p \;=\; Mv \;=\; 5.39 \times 10^{-4}(kg\,m/s)$

$P_A \;=\; v^2 R_{MA} \;=\; [\rho_O/(2\pi c)]\,Va^2 \;=\; 1.112 \times 10^{-3}(W)$

$P_B \;=\; I_B{}^2(R_E + R_{ECOR}) \;=\; 0.274\,(W)$

$P_{CORB} \;=\; I_B{}^2 R_{ECOR} \;=\; 0.031\,(W)$

$P_{EMA} \;=\; I^2 R_{EMA} \;=\; 4.366 \times 10^{-3}(W)$

$PF_E \;=\; P/(E\,I) \;=\; R_{ET}/Z_{ET} \;=\; \cos\phi_E \;=\; 0.9654$

$PL \;=\; (P_{MEC} + P_A)/P_H \;=\; R_{EMA}/(R_E + R_{ECOR}) \;=\; 0.01558$

$PF_{MA} \;=\; (P_{MEC} + P_A)/(F\,v) \;=\; \cos\phi_{MA} \;=\; 0.1363$

$P_{MEC} \;=\; v^2(R_{MEQ} - R_{MA}) \;=\; 3.25 \times 10^{-3}(W)$

$P_H \;=\; I^2(R_E + R_{ECOR}) \;=\; 0.2801(W)$

$P_{MEC} + P_A \;=\; (I\,E_B) \cos\phi_{MA} \;=\; 4.366 \times 10^{-3}(W)$

$pr \;=\; [\rho_O/(2\pi)]\,Va \;=\; 0.2684(\rho\,a)$

$P_{REF} \;=\; E^2/R_{ET} \;=\; I^2 R_{ET} \;=\; E\,I \;=\; 0.320\,watt\;at\,f_E(W)$

$Q \;=\; (E\,I) \sin\phi_E \;=\; I^2 X_{ET} \;=\; 0.0768(VAR)$

$Q_M \;=\; Q_T(PL + 1) \;=\; f_O/(f_2 - f_1) \;=\; 3.5307$

$Q_E \;=\; Q_M/PL \;=\; \omega_O M/R_{MTR} \;=\; 1.0222$

$Q_T \;=\; Q_M/(PL + 1) \;=\; \omega_O M/R_M \;=\; 0.7927$

$R_E \;=\; R_{ET} - (R_{EMA} + R_{ECOR}) \;=\; 5.685(\Omega)$

$R_{ECOR} \;=\; (R_{EMA}/PL) - R_E \;=\; 0.728(\Omega)$

$R_{ET} \;=\; R_E + R_{ECOR} + R_{EMA} \;=\; (PL + 1)\,(R_E + R_{ECOR}) \;=\; 6.513(\Omega)$

$R_{EMA} \;=\; [(Bl)^2/(Z_{MEC} + Z_A)] \cos\phi_{MA} \;=\; 0.100(\Omega)$

$R_G \;=\; assumed\;0(\Omega)$

$R_{IN} \;=\; (Z_{EC} + Z_A) \cos\phi_{MA} - (R_{MS} + 2 R_{MA}) \;=\; 2.2895(N \cdot s/m)$

$R_{MTR} \;=\; (Bl)^2/(R_E + R_{ECOR}) \;=\; 8.3323(N \cdot s/m)$

$R_{MEQ} \;=\; (Z_{MEC} + Z_A) \cos\phi_{MA} \;=\; 9.9234(N \cdot s/m)$

$R_{MA} \;=\; [\rho_O/(2\pi c)]\,\omega^2 S_D{}^2 \;=\; 2.5312(N \cdot s/m^5)$

$R_{MS} \;=\; [2\pi M(f_2 - f_1)] - (2 R_{MA}) \;=\; 2.5861(N \cdot s/m)$

$R_M \;=\; [(Bl)^2/(R_E + R_{ECOR})] + R_{MS} + 2 R_{MA} \;=\; 11.887(N \cdot s/m)$

$S \;=\; E\,I \;=\; I^2 Z_{ET} \;=\; 0.2947(VA)$

$S_D \;=\; \pi a_r{}^2 \;=\; 0.0238(m^2)$

$SPL \;=\; 20 \log [Va\,\rho_O/(2\pi * 2.0^{-5})] \;=\; 82.56(dB)$

$T \;=\; 1/f \;=\; 2.193 \times 10^{-3}(s)$

$U \;=\; v\,S_D \;=\; Va/\omega \;=\; 4.989 \times 10^{-4}(m^3/s)$

$v \;=\; (F/\omega M)\,MRF \;=\; Bl\,I/(Z_{MEC} + Z_A) \;=\; 0.02096(m/s)$

$Va \;=\; a\,S_D \;=\; 1.4293(m^3/s^2)$

$$V_{AT} = \rho_O c^2 C_{MT} S_D{}^2 = 0.02304(m^3)$$
$$W_{ECY} = (E I) \cos \phi_E \, T = 6.239 \times 10^{-4}(J)$$
$$W_{EMACY} = I^2 R_{EMA} \, T = 9.57 \times 10^{-6}(J)$$
$$W_{EHCY} = I^2 (R_E + R_{ECOR}) \, T = 6.143 \times 10^{-4}(J)$$
$$W_{ECLCY} = I^2 R_E T = 5.45 \times 10^{-4}(J)$$
$$W_{ECRCY} = I^2 R_{ECOR} T = 6.97 \times 10^{-5}(J)$$
$$W_{MACY} = (P_{MEC} + P_A) \, T = 9.57 \times 10^{-6}(J)$$
$$X_C = -(k/\omega) = -1.205(mech \, \Omega)$$
$$X_{EMA} = (Bl)^2 (X_{MEC} + X_{MA})/(Z_{MEC} + Z_A)^2 = 0.729(\Omega)$$
$$X_E + X_{ECOR} = (Z_{ET} \sin \phi_E) - X_{EMA} \, 2.487(\Omega)$$
$$X_{ET} = X_{EMA} + X_E + X_{ECOR} = Z_{ET} - j R_{ET} = 1.758(\Omega)$$
$$X_M = \omega M = 73.691(mech \, \Omega)$$
$$X_{MEC} + X_A = (\omega M) - [1/(\omega C_{MT})] = 72.483(mech \, \Omega)$$
$$Z_{EMA} = Blv/I = 0.7331(\Omega)$$
$$Z_{ET} = E/I = R_{ET} + j X_{ET} = 6.7464(\Omega)$$
$$Z_{MEC} + Z_A = \omega M/MRF = Bll/v = 72.886(mech \, \Omega)$$
$$c = 345(m/s)$$
$$\eta = P_A/P = 0.390\%$$
$$\eta_O = P_A/P_{REF} = 0.453\%$$
$$\eta_{EM} = P_{MEC}/P = 0.0114$$
$$\eta_{MA} = P_A/P_{MEC} = 0.342$$
$$\phi_E = arc \cos (P/E I) = arc \cos (R_{ET}/Z_{ET}) = 15.111 \deg .$$
$$\phi_{MA} = arc \cos [(P_{MEC} + P_A)/F v] = arc \cos [R_{MEQ}/(Z_{MEC} + Z_A)] = 82.164 \deg .$$
$$\rho_O = 1.18(kg/m^3)$$
$$\omega = 2 \pi f = 2,865.13(rad/s)$$
$$\omega_O = 2 \pi f_O = 366.37(rad/s)$$
$$\zeta = R_M/(2 M \omega_O) = 1/(2 Q_T) = 0.6308$$

Parameter Tables

Table A1 Parameter values not noted in the text

f	E_B	ϕ_{MA}	P_A	R_{MEQ}	R_{EMA}	R_{MTR}	PL	KE_{MAX}
40.0	0.9348	68.60	0.000318	2.8225	2.5213	9.2594	0.4369	0.0004206
58.3	1.0982	0	0.000934	2.6689	20.0217	9.2186	3.4541	0.0005805
140.0	0.6131	83.31	0.001678	2.2071	0.3285	9.0110	0.0554	0.0001809
228.0	0.3501	84.74	0.001451	3.1734	0.1412	8.8030	0.0233	0.0000590
456.0	0.1532	82.16	0.001112	9.9235	0.1000	8.3323	0.0156	0.0000113

Table A2 Parameter values not noted in the text

f	a	A	V_a	$Z_{MEC} + Z_A$
40.0	32.140	0.0007196	0.7649	7.734
58.3	55.041	0.0005799	1.3099	2.669
140.0	73.779	0.0001348	1.7559	18.948
228.0	68.607	0.0000473	1.6328	34.649
456.0	60.056	0.0000103	1.4293	72.886

Appendix B: New Loudspeaker with Extended Bass[1]

Some of the most popular high-performance loudspeaker systems designed for the home utilize the acoustic-suspension or sealed-box principle. During the past several years, many improvements have been made in the woofer for this type of loud-speaker. New and improved materials and processes, such as polyurethane sur-rounds, vacuum-formed cones, and high-temperature voice coils, have contributed greatly to power-handling ability and smoother upper-range response. The perfor-mance below 500 Hz or so, however, has not been significantly improved, with the low bass response and distortion of the better units of some 20 years ago being essentially equal to those of present day design.

A problem with conventional design is that the laws of physics determine the maximum piston-band efficiency and low-frequency cutoff that can be obtained in a system with a given cabinet volume. If more efficiency is desired, bass response must be sacrificed, and conversely to obtain more bass, it is necessary to sacrifice efficiency.

One known method of circumventing this physical limitation is the negative-spring principle [B1], which is somewhat complex in design. Another is the use of electronic equalization to modify the amplifier response, and this requires additional wattage for any gain achieved. The purpose of this article is to describe a third method that is both simple to implement and that requires no additional wattage from the amplifier.

---

[1] This is a reproduction of the landmark article on W. H. Watkins' patented woofer as it appeared in the December 1974 issue of Audio Magazine.

© The Author(s), under exclusive license to Springer Nature Switzerland AG 2022   109
W. H. Watkins, *Loudspeaker Physics and Forced Vibration*,
https://doi.org/10.1007/978-3-030-91634-3

Back EMF

Since the above problem is related to the motion-induced back EMF generated by the speaker at fundamental resonance, we need to understand its nature. The magnet-voice coil assembly of all dynamic speakers of the type we are discussing is a motor. An induced voltage is produced whenever an electrical conductor moves through a magnetic field; this effect is exhibited by all motors, whether rotary (the conventional type) or linear (such as a speaker) and is termed "back EMF" because its polarity is always opposite to that of the externally applied voltage. So this motion-induced back EMF opposes the amplifier voltage, raises the speaker's impedance, and reduces the current flow through the voice coil.

Since this back EMF is caused by the motion, i.e., velocity of the cone-voice coil, let us see how velocity and thereby back EMF, behaves. In a sealed-box, the enclosed volume of air is not a controlling or limiting factor above the region of resonance. Velocity is low, the air in the box is compressed and rarefied very little, and the air volume reactance is negligible. The output is determined by the cone area, motor, and mass of the cone and voice coil [B2]. Starting at 500 Hz, then, as we come down in frequency, the mass is "easier" to move due to decreasing mass reactance, and velocity doubles with each lower octave. In this area, and down to around 100 Hz, velocity is low, very little back EMF is generated to oppose the amplifier voltage, the back EMF is not a principal factor in determining current flow through the voice coil, and current flow is determined mainly by the d.c. resistance of the voice coil. This is the piston-band of operation where the decreasing air load resistance is offset by the increasing cone velocity, and the speaker output is approximately flat. As we move below 100 Hz or so and approach resonance, the air in the box enters the picture. The mass begins to resonate with the compliance of the air in the box, and velocity increases. When resonance is reached, the moving mass is at any instantaneous moment supplying energy to the air in the box (compressing or rarefying it) due to its kinetic energy, or the air in the box is supplying energy to the moving mass (forcing it in or out) due to its pressure or rarefaction. Thus, the mass and elasticity of the air in the box neutralize each other at resonance and velocity is greatest. With velocity greatest at resonance, the back EMF is greatest, and will raise the impedance by a factor of about 3–6. If the voice coil d.c. resistance is 4 Ω, the back EMF will add about 1 Ω at the resistive point in the 100–500 Hz range for a total of 5 Ω impedance in a typical woofer. Then below 100 Hz the increasing velocity may add 16 Ω at the resonant point due to the large back EMF, and the total impedance will be 20 Ω. This is shown in Fig. B.1. Since the two points are resistive, we can calculate the power input to the voice coil by

$$\text{Watts} = \frac{E^2}{R}$$

Where E = Voltage from the amplifier in volts

R = Resistance in ohms.

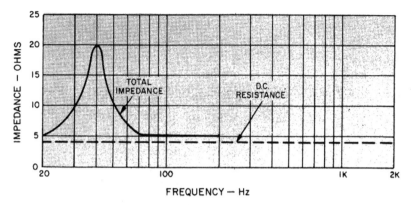

Fig. B.1 Typical impedance curve. The total impedance is the d.c. resistance plus the reactance due to the back EMF

With E essentially constant, and with a voltage of 10 volts, the voice coil is absorbing 20 watts at the point above resonance and only 5 watts at resonance. Since a speaker is more efficient at resonance, this condition produces flat response in conventional design. It is clear however, that while the back EMF was not a principal factor in determining current flow and wattage through the voice coil above resonance, it has become the dominant factor at resonance and has severely limited the wattage intake of the speaker. In fact, if the power intake at resonance could be brought up to 20 watts and electro-mechanical losses were negligible, the relative bass output would increase by four times, that is, it would be elevated 6 dB above the higher-frequency output. If the output at these higher frequencies above resonance could then be increased, the net result would be an over-all higher output.

Dominant Parameters

While the device to be decribed may be characterized in terms of Q_T (shape of curve at resonance), it is believed that the approach used here offers a better "feel" for the principle involved. Although some approximations are necessary due to different factors in dominant control in given areas, the pattern of behavior is clearly shown [B3].

The reasonable assumptions made are that the inductance of the voice coil below 500 Hz is negligible, that the amplifier source resistance is negligible (a high damping factor), and that good design is used throughout, and in particular in keeping proper control of the mechanical and acoustic resistive elements [B4].

Since acoustic power output is proportional to velocity squared times the radiation resistance, and back EMF is also proportional to velocity, we will work mainly with this term. In the region above resonance and below 500 Hz, say at 200 Hz, where the cone of a 12-in. woofer is vibrating as a piston, the cone velocity is

$$V = \frac{F}{Zm} \tag{B.1}$$

Where F = Force applied to the voice coil

Z_m = Mechanical impedance
And F = Bli

Where

$$B = \text{Flux density in the air gap} \tag{B.2}$$

I = Length of voice coil wire in the magnetic field

i = Current in the voice coil

So

$$V = \frac{Bli}{Zm} \tag{B.3}$$

To find the current, the total electrical impedance [B5] is

$$Z_{ET} = Rc + \frac{B^2 l^2}{Zm} \tag{B.4}$$

Where Rc = Electrical impedance of the voice coil with no motion.
Now, above resonance (Fr), Rc dominates, and the current is

$$i \approx \frac{Eg}{Rc} \tag{B.5}$$

Where Eg is the amplifier voltage.
Substituting for i in Eq. (B.3), velocity above resonance

$$V > Fr \approx \frac{Bl \; \frac{Eg}{Rc}}{Zm} \approx \frac{Bl}{Rc} \frac{Eg}{Zm} \tag{B.6}$$

Then at resonance in Eq. (B.4), $\frac{B^2 l^2}{Zm}$, the motional impedance term which is responsible for back EMF dominates, and the current is

$$i \approx \frac{Eg}{B^2 l^2 / Zm} \approx \frac{Eg \, Zm}{B^2 l^2} \tag{B.7}$$

Substituting for i in Eq. (B.3), velocity at resonance

$$V_{Fr} \approx \frac{Bl\left(\frac{Eg}{B^2l^2} \frac{Zm}{}\right)}{Zm} \tag{B.8}$$

And

$$V_{Fr} \approx \frac{Eg}{Bl} \tag{B.9}$$

Zm varies with frequency, is a minimum at Fr, and Eq. (B.4) shows that this low mechanical impedance at Fr is reflected back into the electrical circuit by the back EMF as a high impedance. So, as noted before, the voice coil does not absorb nearly as much power at Fr as it can above and below. Equation (B.9) shows that motor (Bl) is mainly in control of output at Fr. Since back EMF is proportional to B l, a higher Bl gives more back EMF at resonance, and the and acoustic output go down. Equations (B.6) and (B.9) show that for a given speaker, that is, a given area of cone, mass, and volume there will be an optimum value, neither too small or too large, of Bl product that will give the flattest response.

The See-Saw Effect

Now let's look at the performance of a loudspeaker system in which the sealed volume, cone area, and mass are kept constant and plot the results of changing motor strength. In Fig. B.2, as shown by Eq. (B.6), in the 70–500 Hz range if Bl is increased output will increase, if it is decreased output will decrease. However, around resonance the reverse happens. As shown by Eq. (B.9), increasing Bl decreases output and decreasing Bl increases output at resonance. The results shown in Fig. B.2 are very similar to what would be obtained by denormalizing the standard Qt curves of .5 and 2 in Fig. B.3 [B6]. The curves of Fig. B.3 are generally referred to in discussing bass response. However, it must be noted that they have all been normalized with respect to actual flat band efficiency of each; i.e., the O dB point on the vertical axis is redefined for each speaker, regardless of actual efficiency, This normalization process does however, have a disadvantage, in that when adjusting the Bl product, it does not clearly show that a highly damped speaker will be more efficient above resonance than a less damped unit which peaks at resonance.

From Fig. B.2 it can be seen that for a given speaker system, and where flat response is desired, the motor must be of the proper strength. Balancing the bass response to the flat band is much like balancing a see-saw, with the pivot point at 70 Hz or so. If the motor is too small, efficiency is low, and there is a bump in the bass. If it is too large, efficiency is high, but the bass response is down. This also shows that purchasing the speaker system with the larger magnet (it is often assumed that the speaker with the largest magnet is best) could buy one a speaker with less

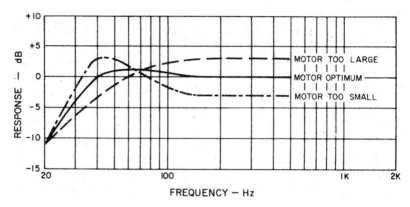

Fig. B.2 Response for different values of motor strength in the region of fundamental resonance with conventional design

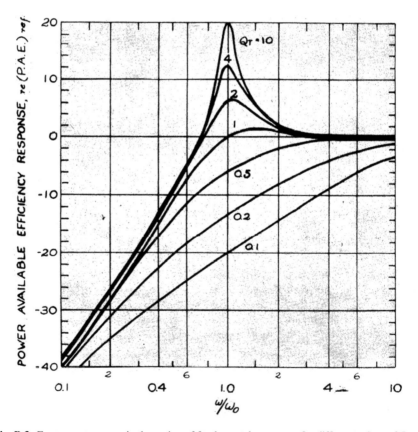

Fig. B.3 Frequency response in the region of fundamental resonance for different values of Q_T

than optimum bass response. By juggling motor then, there is an inevitable trade-off between bass response and efficiency in the flat band.

With a given size enclosure, the foregoing discussion shows the limitations which must prevail. Efficiency in the 70–500 Hz midrange area is tied to efficiency in the bass range, and the see-saw effect prevails. When driven by modern-day solid-state amplifiers with their high damping factors, measurements and/or reference to reliable test report curves will show that many of the better speaker systems will begin to fall off in bass response above resonance. This is due to the motor having been increased (the "see-saw" tilted up in favor of the midrange) in order to obtain reasonable efficiency, i.e., a total power output above resonance of about 80 dB SPL (sound pressure level), re to .0002 microbar at 1 meter distance for 1 watt of electrical input. So in the final analysis, the crux of the problem is due to the see-saw effect of varying motor strength. For good efficiency above resonance, we need a large motor; for good bass efficiency, we need a smaller motor.

Dual Motors

A solution is to provide the speaker with two motors This would cut loose the see-saw effect, and we would be free to adjust upper range and bass separately, and without one affecting the other. The usual practice in adjusting motor strength is to vary the magnetic strength, that is, the B in the Bl product. To construct a speaker with two different magnetic field densities to drive the same cone would be both expensive and difficult to manufacture.

Let's look at the Bl product in a different way. In a conventional speaker, it is constant with frequency. Suppose we could in effect make I vary with frequency in such a manner that a lower value of Bl in one frequency range would not affect a higher Bl product in another range, and vice versa. Figure B.4 shows a simple method of doing this. A second voice coil is wound over or under the conventional voice coil and is driven by a series LC resonant circuit adjusted to resonate at the same frequency as the fundamental mechanical-acoustical resonant frequency of the woofer. With proper design of the inductor L and capacitor C, the LC circuit presents almost zero impedance at resonance, and a sufficiently high impedance one octave either side of resonance to effectively remove voice coil 2 from the circuit. Thus I of voice coil 2 can be adjusted to eliminate the high value of motion-induced back EMF at fundamental resonance, a lower impedance path is provided at Fr for current flow, and bass response at Fr can be adjusted at will and independently of midrange response above Fr and response below Fr. Taking our example used before, this will allow the impedance to be lowered to say 5 Ω (compared to 20-Ω before) at resonance, and the wattage intake can be brought up if needed to equal that above resonance (20 watts). At this point it should be noted that no attempt is made to violate the laws of physics. A second motor is added that generates less back EMF and offers a lower impedance to the amplifier, allowing more wattage intake to the speaker in the region of a resonance, and providing more bass response. We are

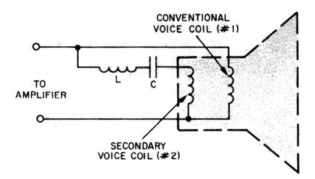

Fig. B.4 Circuit for a dual-motor woofer. L and C are adjusted to resonate at the same frequency as the fundamental acoustic resonance of the woofer. The tuned bandpass circuit formed by L and C allow the secondary voice coil to operate only in the region of resonance

simply using the wattage in a given amplifier that was **there all the time,** but not being used in the bass range.

The additional cost of building a woofer of this type over a conventional woofer is a pleasant surprise. A voice coil gap of twice the width is not required (a wider gap requires a larger magnet to obtain the same flux density) to accommodate the extra voice coil. The outer and inner clearance spaces are the same as usual, and since a single layer coil with a notch in the top plate for the return end of the coil has proved satisfactory, the gap width need only be increased by about 25%. Any extra money put into a larger magnet can be utilized to provide benefits that can be put to good use.

Once it is realized that the motor can be increased to the point of gap saturation with the attendant overdamping of bass compensated for by the second voice coil, the dividends of the design are apparent (as the flux density is increased, the impedance at resonance is increased, and fewer turns may be used in the second voice coil without its impedance falling to an unacceptable value). More overall efficiency, a smaller enclosure, extended bass as noted, or a combination of these may be achieved, according, to how the parameters are chosen, and with the usual tradeoffs in these areas minimized. For the same shape of curve and volume of box the maximum realizable efficiency gain is calculated to be slightly over 3 dB.

Several variations of the principle involved were tried, but more research is needed in some areas. While using the circuit of Fig. B.4, if the main voice coil is disconnected at resonance, a further increase in acoustic output occurs. The measured EMF across it is higher than the amplifier drive voltage by about the same ratio as the ratio of turns of wire in the two voice coils. Inserting a parallel LC circuit in series with the main voice coil allows the increase to be utilized.[2] The Q of the LC circuits are controlled in a straightforward manner by the LC ratio. Mutual coupling of the two inductors in the two LC circuits can also be accomplished.

---

[2]A capacitor in series with the main voice coil was used in the test speaker for this article.

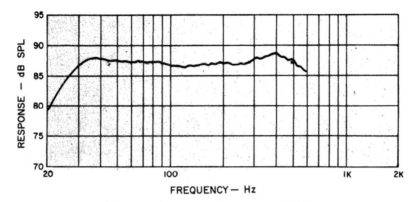

Fig. B.5 Frequency response of a dual-motor (1 watt into rated impedance of 4 Ω), mike at 1 m on-axis, 180° radiation angle, and free-field conditions. SPL ref. is 0.0002 μ Bar

Performance

Figure B.5 shows the frequency response of a woofer in which a gain of 5½ dB at 30 Hz was achieved. In all the tests the speaker was sunk in the ground with the front edge flush with the ground surface, and well away from any buildings (180° solid angle, free-field conditions), and the microphone was at 1-m distance, on-axis with the geometrie center of the speaker.

The following test equipment was used for the performance figures shown:

Hewlett-Packard 200-CD audio oscillator, Hewlett-Packard 400-H A.C. vacuum-tube voltmeter, Hewlett-Packard 120-B oscilloscope, Hewlett-Packard 330-B distortion analyzer, Crown DC-300A amplifier, B & W 410 a.c. vacuum-tube voltmeter & distortion analyzer, General Radio 1390-B random noise generator with P2 pink noise filter, General Radio 1564-A sound and vibration analyzer. General Radio 1962 ½-inch condenser microphone, General Radio 1560-P42 microphone preamplifier. General Radio 1562-A sound level calibrator, custom-built a.c. impedance bridge, and Eico 377 square wave generator.

The woofer is a 12-in. unit (9½-in. actual cone diameter), mounted in a box of two-cubic-foot net internal volume, and the box filled with a suitable amount of fiberglass.

In constructing this unit, primary emphasis was placed on extending the bass, slight emphasis on efficiency, and none on reducing enclosure volume. The reason for this is that the writers experience has shown that, other being equal, the loudspeaker listener will usually prefer the speaker with the cleanest and most extended bass response.

Taking 87 dB SPL (reference .0002 microbar) produced with 1 watt input (2 volts into a rated impedance of 4 Ω) as reference, the response is flat to 30 Hz, and down only 3 dB at 25 Hz. The response is flat ±1½ dB from 27 to 600 Hz. It has been stated [B7] that a range of 30–15,000 Hz is required to reproduce orchestral music

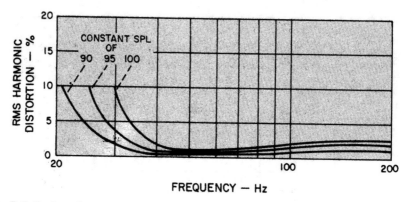

Fig. B.6 Total rms harmonic distortion with free-field conditions as noted previously. The sound pressure level was held constant for each curve, and the frequency lowered until the distortion reached 10%

with no discernible frequency discrimination. This is in agreement with the lowest musical tones generally found in recorded program material of the piano, double-bass viol, harp, organ, and drums. The curve of Fig. B.5 is unusual, considering the 2-cubic-foot volume, the overall efficiency, it being flat to 30 Hz, and that no amplifier equalization is used.

Total rms harmonic distortion is shown in Fig. B.6. Note that the curves are for constant sound pressure levels, not constant wattage input. A specification showing low distortion down to 30 Hz with a constant wattage input is not very meaningful, if the acoustic output of the speaker is down 10 dB at this frequency.

Free-field specifications can be taken as statements of fact and offer data for direct comparison, but an assessment of performance in the listening room—even an "average" room—involves many variables, and becomes a difficult task. Furthermore, no standard has been adopted on either the maximum SPL needed or the limit on distortion. These are subjective matters, and it is therefore difficult to assign rigorous figures. The effect of boundary reflections on the response in the listening area is also of importance. Much work is now being done in these areas, and we can expect more useful information to become available.

Nevertheless, it is desirable to assess the performance of a new device in terms of practical use, and an attempt will be made to present the low bass capability in a manner that will allow the reader to make his own evaluation. We need to decide on the lowest frequency to be reproduced, the SPL it is to be reproduced at, and the amount of distortion to be tolerated. Let's set the low frequency at 30 Hz as mentioned earlier, the SPL at 100 dB (extended listening at levels much higher than this can cause hearing impairment), and the distortion limit on the order of 5% (this level of distortion is tolerated in loudspeakers designed for demanding use at higher frequencies and lower volume levels than we are talking about here). At 30 Hz and radiating into 180° free-field, the distortion at 97 dB SPL at one meter is 5.2%. Now if we move the speaker into the listening room, place it against a wall and

Fig. B.7 Acoustic output of
the woofer at 30 Hz and
97 dB SPL. Free-field
conditions as noted
previously. (Oscilloscope of
microphone output)

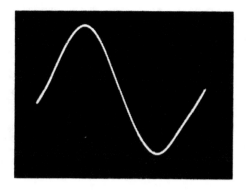

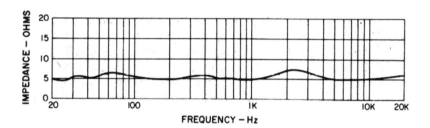

Fig. B.8 Impedance of a system using dual-motor woofer

add another unit for stereo use, the SPL will double and will be 100 dB. In an
"average" room of 2500 cubic feet, and at this low frequency, the reverberant field
SPL should equal the direct field SPL at about one meter [B8], so the 100 dB SPL
should prevail in the listening area. On this basis, we have a woofer in a two-cubic-
foot box, flat to 30 Hz, with a stereo pair producing 100 dB SPL at 30 Hz in the
listening area at a distortion level of only 5.2%, and with a power input to each of
only 10 watts. If the SPL is lowered to 95 dB, the harmonic distortion drops to 2.3%,
and at 90 dB it is 1%. This 5.2% low-bass harmonic distortion in a woofer still allows
for a surprisingly clean waveform, if clipping of the peaks and sharp aberrations are
missing. Figure B.7 is an oscilloscope photo of the acoustic output at 30 Hz at the
SPL producing 5.2% distortion.

The impedance curve for a complete system is shown in Fig. B.8. In one
prototype, the curve was ruler flat below 200 Hz, however, the best overall results
were obtained with some deviation from this ideal, and the curve shown is the one
obtained from the design giving the other data presented in this article. The minimum is 4½ Ω at 25 Hz, the maximum is 7 Ω at about 2200 Hz, and a 4-Ω rating is
appropriate. It does not fall under 5 Ω below 20 Hz, and above 20,000 Hz, there is a
gradual rise due to the usual inductance of the tweeter voice coil. The conventional
bump at bass resonance is missing, due to the effect of the second voice coil
increasing the power intake in this region.

Fig. B.9 Acoustic output of
the woofer in response to the
step-front of a low-
frequency square wave

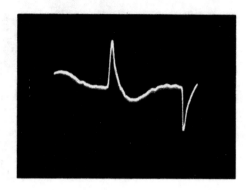

Transient response refers to the ability of the woofer to follow the starting and stopping (attack and decay) of sudden changes in the electrical signal. In a woofer, this is a function of the system Q. It is generally considered that a Q of one offers the flattest response with a minimum of ringing, and the response curve of Fig. B.5 conforms to this shape. Since the woofer is driven by a series resonant LC circuit, we need to see if the transient response follows the rules. It has been shown [B9] that a woofer in a sealed-box with a Q of one will have about 33% overshoot to a step-signal, and will decay within one cycle. Figure B.9 is an oscilloscope photo of the wave-form of the acoustic output in response to the step-front of low-frequency square wave. The overshoot past the neutral or start position is exactly that expected with a Q of one, and shows that the woofer is properly damped both acoustically and electrically. This test was made indoors, and the small ripples in the waveform are due to room boundary reflections.

System Performance

The frequency response of a complete system consisting of a 12-in. woofer, 4½-in. mid-range, and 1-in. dome tweeter is shown in Fig. B.10. One-third-octave bands of random noise were used to produce the curves. The dot-dash line is the on-axis response, and the solid line is the average of five frontal curves; one on-axis, and four at 45°, with the microphone at the sides and ends, one meter away from and aimed at the geometric center of the box. Taking 85½ dB SPL as reference, the lower curve (a good indication of the overall response), shows the system to be ±3 dB from 23 through 20,000 Hz. The system's impedance curve is that of Fig. B.8. Although the output of the midrange driver needs to be increased slightly, the response is relatively smooth.

In listening comparison using program material with deep bass content, the low bass is simply "there." The lowest notes are reproduced cleanly at equal volume, and the temptation to turn up the bass control is missing. The low notes have an "easy

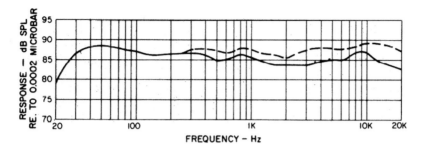

Fig. B.10 Frequency response of a three-way system using the dual-motor woofer with one-third-octave random noise. Speaker radiating into 180° free-field conditions as noted, with one-watt input

unstrained" sound, which is attributed to the linearity and low phase shift of the load presented to the amplifier.

In summary, the design does not involve trade-offs in areas of performance, requires no additional amplifier power or equalizer, has the advantage of simplicity of construction, and offers an improvement that is readily discernible on listening.

A patent application has been allowed for the loudspeaker system described in this article.

Glossary and Symbols[3]

A	Amplitude (distance from equilibrium to an end point of motion)
a	Acceleration (time rate of change of velocity)
a_r	Cone radius including half of the surround
Bl	Motor force constant (motor strength)
$(Bl)^2$	Transformation factor
BW	Bandwidth at mechanical resonance
C_{MT}	System mechanical compliance (assumed constant with frequency)
E	Applied *RMS* sinusoidal voltage
E_B	Generated ("open circuit") back emf or counter voltage
E_{BEFF}	Effective component of generated back emf
EBP	Efficiency bandwidth product
E_{DR}	Drive voltage (portion of applied voltage driving the mechanical plus acoustic load)
E_H	Portion of applied voltage generating heat in the coil and core
f	Natural frequency
f_3	3-dB down point in the low end frequency response (with reference to the piston-band)
f_E	Electrical resonant frequency (frequency of minimum impedance above mechanical resonance where voltage and current are in phase and EI is real power)
f_O	Mechanical resonant frequency
F_{ST}	Electromagnetic starting force at the instant of start of motion

---

[3] Subscripts denoting electrical, mechanical, acoustic, and mechanical-plus-acoustic quantities are denoted respectively as subscripts $_{E, M, A,}$ and$_{MA}$.. A basic equation for each parameter is given in the Appendix, along with the magnitude at 456 Hz for a reference driver. Magnitudes at the basic test frequencies not noted in tables are also given in the Appendix.

© The Author(s), under exclusive license to Springer Nature Switzerland AG 2022
W. H. Watkins, *Loudspeaker Physics and Forced Vibration*,
https://doi.org/10.1007/978-3-030-91634-3

F_{MO}	Total applied electromagnetic force during steady-state motion
F	Portion of applied force F_{MO} driving the mechanical and acoustic load (driving force)
F_{ED}	Magnitude of the portion of applied force F_{MO} overcoming motor electromagnetic damping force
F_{DMTR}	Magnitude of effective motor electromagnetic damping force
F_{MB}	Magnitude of generated motor electromagnetic damping force
f_{ZMIN}	Equivalent to f_E
I	*RMS* current with steady-state motion
I_B	*RMS* current with the coil blocked for no motion
k	Stiffness of the suspension system, including air in the enclosure
$KE_{(MAX)}$	Maximum kinetic energy of the moving mass
L_E	Inductance of coil and core plus that due to motion as reflected into the electrical circuit
L_G	Inductance of source, assumed negligible
M	Mass of the mechanical moving assembly plus the air load on both sides of the cone (assumed constant with frequency)
M_{MD}	Mass of the mechanical moving assembly only
MRF	Mechanical response function (ratio of inertial force Ma to electromagnetic driving force F, or ratio of mass reactance to equivalent mechanical resistance)
P	True electrical power delivered by the source during steady-state motion
p	Momentum
P_A	Acoustic power radiated on the front side of the cone
P_B	True electrical power delivered by the source with coil blocked for no motion
P_{CORB}	Power dissipated into the core
P_{EMA}	Power from the source used to overcome the retarding forces due to motion
PF_E	Electrical power factor (cosine of the phase angle between voltage and current)
PL	Power-to-loss ratio (ratio of mechanical plus acoustic power to power lost as heat in the coil and core in the energy conversion process)
PF_{MA}	Mechanical power factor (cosine of the phase angle between force and velocity)
P_{MEC}	Mechanical power
P_H	Power lost as heat in coil and core (copper loss + loss in core)
$P_{MEC} + P_A$	Mechanical plus acoustic power
pr	Air pressure at one meter distance, on axis, into an angle of 180 degrees (free field)
P_{REF}	Electrical power dissipated during steady-state motion at f_E
Q	Quadrature (reactive) electrical power
Q_M	Q of driver considering non-electrical resistances only

Q_E	Q of driver considering electrical resistances only
Q_T	Total Q of driver, considering all system resistances
R_E	DC resistance of the voice coil
R_{ECOR}	Resistive component of the motor metal works (mainly the pole piece) due to eddy current flow
R_{ET}	Total resistance seen by the source
R_{EMA}	Effective electrical resistance due to the real part of the mechanical and acoustic impedance as reflected into the electrical circuit
R_G	Resistance of the source (assumed to be zero)
R_{IN}	Resistance due to inertia
R_{MTR}	Electromagnetic damping due to the motor
R_{MEQ}	Equivalent mechanical resistance
R_{MA}	Mechanical resistance of the air load on the front side of the cone
R_{MS}	Mechanical resistance of the suspensions (assumed constant with frequency)
R_M	Equivalent mechanical resistance at the instant of start of motion
S	Apparent or potential electrical power
S_D	Cone area (including half of the surround)
SPL	Sound pressure level, on axis, at one meter distance, and radiating into 180 degrees (free field)
T	Period (time taken to complete one cycle)
U	Volume velocity
v	True RMS velocity of the moving mass assembly
Va	Volume acceleration (time rate of change of volume velocity)
V_{AT}	System compliance expressed as an equivalent volume of air
W_{ECY}	Total work done per cycle by the electrical source
W_{EMACY}	Work done per cycle by the electrical source on the mechanical and acoustic load
W_{EHCY}	Work done per cycle by the electrical source creating heat in the coil and core
W_{ECLCY}	Work done per cycle by the electrical source heating the coil
W_{ECRCY}	Work done per cycle by the electrical source heating the core
W_{MACY}	Work done per cycle on the mechanical and acoustic side expressed in mechanical and acoustic terms
X_C	Compliance or stiffness reactance
X_{EMA}	Reactance due to motion as reflected into the electrical circuit (negative above mechanical resonance)
$X_E + X_{ECOR}$	Reactance of the coil and core
X_{ET}	Total electrical reactance
X_M	Mass reactance
$X_{MEC} + X_A$	Mechanical plus acoustic reactance
Z_{EMA}	Electrical impedance due to motion as reflected into the electrical circuit
Z_{ET}	Total electrical impedance

$Z_{MEC} + Z_A$	Mechanical plus acoustic impedance
c	Velocity of sound through the air
η	True efficiency
η_O	True reference efficiency at f_E
η_{EM}	Electrical to mechanical efficiency
η_{MA}	Mechanical to acoustic efficiency
ϕ_E	Electrical phase angle (phase angle between voltage and current)
ϕ_{MA}	Mechanical plus acoustic phase angle (phase angle between force and velocity)
ρ_O	Density of the air
ω	Radian frequency
ω_O	Radian frequency at mechanical resonance
ζ	Total system damping ratio

References

1. A.N. Thiele, Loudspeakers in vented boxes, Parts I and II. J. Audio Eng. Soc. **19**, 382–392, pp. 478–483 (1971)
2. I.M. Dash, An Equivalent Circuit Model for the Moving Coil Loudspeaker, Honours Thesis, School of Electrical Engineering, University of Sydney, Australia, 1982
3. J. Vanderkooy, A model of loudspeaker impedance incorporating eddy currents in the pole structure. J. Audio Eng. Soc. **37**, 119–128 (1989)
4. J.R. Wright, An empirical model for loudspeaker motor impedance. J. Audio Eng. Soc. **38**, 749–754 (1990)
5. W. Marshall Leach Jr., Loudspeaker voice-coil inductance losses: circuit models, parameter estimation, and effect on frequency response. J. Audio Eng. Soc. **50**(6), 442–450 (2002)
6. L.L. Beranek, *Acoustics (Eq. 1.11)*. (McGraw-Hill, London, 1954), p. 11. Reprinted by the Acoustical Society of Americae, Woodbury, NY, (1996)
7. L.E. Kinsler, A.R. Frey, A.B. Coppens, J.V. Sanders, *Fundamentals of Acoustics (Eq. 14.51b)*, 3rd edn. (Wiley, New York, 1982), p. 361
8. W.T. Thomson, *Mechanical Vibrations (Eq. 63.3)*, 2nd edn. (Prentice-Hall, Englewood Cliffs, NJ, 1953), p. 234
9. P.M. Morse, *Vibration and Sound* (American Institute of Physics for The Acoustical Society of America, 1976), p. 35
10. M. Dodd, W. Klippel, J. Oclee-Brown, Voice coil impedance as a function of frequency and displacement. J. Audio Eng. Soc., Preprint Number 6178, Convention 117, Fig. 31, (2004)
11. L.E. Kinsler, A.R. Frey, A.B. Coppens, J.V. Sanders, *Fundamentals of Acoustics (Eq. 14.51b)*, 3rd edn. (Wiley, New York, 1982), p. 361, 366
12. E.M. Villchur, *Handbook of Sound Reproduction* (Radio Magazines Inc., Mineola, NY, 1957), p. 135
13. J. King, Loudspeaker voice coils. J. Audio Eng. Soc. **18**(1), 34–43 (1970)
14. L.L. Beranek, *Acoustics (Eq. 7.5)* (McGraw-Hill, London, 1954), p. 188. Reprinted by the Acoustical Society of America, Woodbury, N.Y., (1996)
15. R.H. Small, Direct-radiator loudspeaker system analysis. J. Audio Eng. Soc. **20**(5), 383–395 (1972)
16. L.E. Kinsler et al., *Fundamentals of Acoustics (Eq. 14.54a)*, 3rd edn. (Wiley, New York, 1982), p. 361
17. H. Kloss, *Loudspeaker Design* (Audio Magazine, 1971), pp. 30–32, 56
18. R.H. Small, Closed-box loudspeaker systems, Part 1: Analysis. J. Audio Eng. Soc. **20**(10), 798–808 (1972)

© The Author(s), under exclusive license to Springer Nature Switzerland AG 2022
W. H. Watkins, *Loudspeaker Physics and Forced Vibration*,
https://doi.org/10.1007/978-3-030-91634-3

19. A.E. Fitzgerald, C. Kingsley Jr., S.D. Umans, *Electrical Machinery*, 6th edn. (McGraw-Hill, New York, 2003), p. 118, 152
20. H.F. Olson, *Acoustical Engineering (Eq. 6.3)* (D. Van Nostrand Co. Inc., Princeton, NJ, 1957), p. 126
21. W. Marshall Leach Jr., Electroacoustics & Audio Amplifier Design, in *Kendall/Hunt Pub*, (Dubuque, Iowa, 1998), p. 54, 115
22. A.W. Smith, J.N. Cooper, *ELEMENTS OF PHYSICS*, 8th edn. (McGraw-Hill, New York, 1972), p. 515
23. C.E. Swartz, *Phenomenal Physics* (Wiley, New York, 1981), p. 625
24. J.B. Calvert, Univ. of Denver at Denver, Colorado. www.du.edu/~jcalvert/tech/elmotors.htm
25. L.L. Beranek, *Acoustics (Eq. 7.9)* (McGraw-Hill, London, 1954), p. 189. Reprinted by the Acoustical Society of America, Woodbury, N.Y., (1996)
26. D.B. Keele Jr., Low-frequency loudspeaker assessment by nearfield sound-pressure measurement. J. Audio Eng. Soc. **22**(3), 154–162 (1974)
27. H.F. Olson, *Acoustical Engineering (Eq. 6.11)* (D. Van Nostrand Co. Inc., Princeton, NJ, 1957), p. 132
28. L.L. Beranek, *Acoustics (Eq. 7.1)* (McGraw-Hill, London, 1954), p. 188. Reprinted by the Acoustical Society of America, Woodbury, N.Y., (1996)
29. W.H. Watkins, *New Loudspeaker with Extended Bass* (Audio Magazine, 1974), pp. 38–46
30. L.E. Kinsler et al., *Fundamentals of Acoustics (Eq. 14.47)*, 3rd edn. (Wiley, New York, 1982), p. 359
31. L.L. Beranek, *Acoustics (Eq. 8.39)* (McGraw-Hill, London, 1954), p. 230. Reprinted by the Acoustical Society of America, Woodbury, N.Y., (1996)
32. W. Marshall Leach Jr., Electroacoustics & Audio Amplifier Design, in *Kendall/Hunt Pub*, (Dubuque, Iowa, 1998), p. 125
33. L.L. Beranek, *Acoustics* (McGraw-Hill, London, 1954), p. 226. Reprinted by the Acoustical Society of America, Woodbury, N.Y., (1996)
B1. T. Matzuk, Improvement of low-frequency response in small loudspeaker systems by means of the stabilized negative-spring principle. J. Acoust. Soc. Am. **49**(5, Part 1), 1362 (1971)
B2. H. Kloss, *Loudspeaker Design* (Audio, 1971), p. 30
B3. J.D. Tillman, unpublished class notes (Electroacoustics-EE.4570) Electrical Eng. Dept., University of Tenn., Knoxville, Tenn
B4. R.F. Allison, Low-frequency response and efficiency relationships in direct-radiator loudspeaker systems. J. Audio Eng. Soc. **13** (1965)
B5. H.F. Olson, *Acoustical Engineering* (D. Van Nostrand, 1957), p. 130
B6. L.L. Beranek, *Acoustics* (McGraw-Hill, 1954), p. 226
B7. H.F. Olson, *Psychology of Sound Reproduction* (Audio, 1972), p. 20
B8. R.F. Allison, R. Berkovitz, The sound field in home listening rooms. J. Audio Eng. Soc. **20**, 459 (1972)
B9. R.H. Small, Closed-box loudspeaker systems. J. Audio Eng. Soc. **20**, 798 (1972)

Index

A
Acceleration, v, 2, 8, 29, 38, 39, 42, 75, 85, 123
Amplitude, v, 2, 59, 77, 83, 86–87, 123

B
Back emf (counter voltage), v, 1, 2, 7, 12, 25, 37–42, 51, 52, 62, 88, 91–94, 96, 98, 100, 103, 110–113, 115, 123
Bandwidth (at mechanical resonance), 10, 19, 29, 52, 57, 58, 82–83, 92, 95, 123, 126

C
Compliance
  assumed constant with frequency, 123
  mechanical, 8, 10, 20, 34, 45, 65, 82, 123
  system, 8, 34, 82, 92, 123, 125
Cone area (including half of the surround), 8, 69, 96, 98, 99, 110, 113, 123, 125
Cone radius, 123
Current
  blocked-coil, 14–17
  steady-state, 2, 63

D
Density
  air, 126

E
Effective electrical resistance, 10, 38, 125

Efficiency
Efficiency
  true, 1, 2, 8, 27, 30, 32–34, 126
Efficiency-bandwidth product, 57, 123
Electrical-to-mechanical efficiency ratio, 123
Electromagnetic damping, 1, 62–63, 66, 83, 124, 125
Equivalent mechanical resistance, 46–47, 62, 75, 124, 125

F
Force, v, vi, 1–3, 7, 12, 29, 37–42, 46, 50–52, 61–66, 68, 77, 86, 88, 92, 94, 112, 123, 124, 126
Frequency
  natural, xv, 17, 21, 25, 28, 40, 55, 79, 107, 108
  radian, 126

I
Impedance
  acoustic, v, 1, 2, 9, 11, 13, 15, 18, 21, 22, 32, 37, 38, 45, 93, 95, 96, 98, 100, 101, 113, 125
  electrical, v, 1, 2, 9, 11, 18, 19, 21, 22, 32, 37, 38, 91, 94–96, 98, 112, 113, 125
  mechanical, v, 1, 2, 9, 11, 13, 14, 21, 22, 32, 37, 38, 45, 61, 95, 96, 98, 102, 113
Inductance, 2, 7, 10, 13–15, 21–23, 88, 89, 93, 101, 111, 119, 124
Inductance (coil and core), 2, 10, 13, 14, 18, 88, 93, 102, 111, 119

Printed in the United States
by Baker & Taylor Publisher Services